o a d e m e c u

T0261918

Vital Signs
and Resuscitation

Joseph V. Stewart, M.D.
Chairman, Department of Emergency Medicine
Palmetto Baptist Medical Center
Columbia, South Carolina

Adjunct Lecturer, Gross Anatomy
South Carolina School of Medicine
Columbia, South Carolina

Assistant Professor of Medicine
The Chicago Medical School
North Chicago, Illinois

Former Professor of Anatomy and Physiology
Triton College
Rivergrove, Illinois

CRC Press
Taylor & Francis Group
Boca Raton London New York

CRC Press is an imprint of the
Taylor & Francis Group, an **informa** business

VADEMECUM
Vital Signs and Resuscitation

First published 2003 by Landes Bioscience

Published 2018 by CRC Press
Taylor & Francis Group
6000 Broken Sound Parkway NW, Suite 300
Boca Raton, FL 33487-2742

ISBN 13: 978-1-57059-671-1 (pbk)

This book contains information obtained from authentic and highly regarded sources. Reasonable efforts have been made to publish reliable data and information, but the author and publisher cannot assume responsibility for the validity of all materials or the consequences of their use. The authors and publishers have attempted to trace the copyright holders of all material reproduced in this publication and apologize to copyright holders if permission to publish in this form has not been obtained. If any copyright material has not been acknowledged please write and let us know so we may rectify in any future reprint.

Visit the Taylor & Francis Web site at
http://www.taylorandfrancis.com

and the CRC Press Web site at
http://www.crcpress.com

Library of Congress Cataloging-in-Publication Data

Stewart, Joseph V., 1931-
      Vital signs and resuscitation / Joseph V. Stewart.
        p. ; cm. -- (Vademecum)
      Includes bibliographical references and index.
      ISBN 1-57059-671-9 (spiral)
       1. Vital signs--Handbooks, manuals, etc. 2. Physical diagnosis
      --Handbooks, manuals, etc. 3. Resuscitation--Handbooks,
      manuals, etc. I. Title II. Series.
      [DNLM: 1. Physical Examination--methods. 2. Blood
      Pressure Determination. 3. Body Temperature. 4. Pulse. 5.
      Respiration. 6. Resuscitation. WB 205 S849v 2001]
      RC76 .S745 2003
      616.07'54--dc21

While the authors, editors, sponsor and publisher believe that drug selection and dosage and the specifications and usage of equipment and devices, as set forth in this book, are in accord with current recommendations and practice at the time of publication, they make no warranty, expressed or implied, with respect to material described in this book. In view of the ongoing research, equipment development, changes in governmental regulations and the rapid accumulation of information relating to the biomedical sciences, the reader is urged to carefully review and evaluate the information provided herein.

Dedication

To Judith, Holly, and Margaret

Contents

Preface.. vii

1. History of the Vital Signs 1
The Thermometer and Temperature 1
Heart Rate and Pulse 6
Respiration 9
Blood Pressure 11
Origin of the Term "Vital Signs" 14
Level of Consciousness 16

2. Vital Sign #1: Temperature 20
Heat Production and Loss 20
Perception of and Reaction to Hot and Cold 20
Acclimatization 23
Body Temperature 23
Methods of Temperature Measurement 23
High Temperature 26
Fever 26
Heat Stroke 27
Heat Exhaustion 29
Uncommon Heat Illnesses 29
Low Temperature (Hypothermia) 30
Infants and the Elderly 31
Practical Points 32

3. Vital Sign #2: Heart Rate/Pulse 34
The Heart: Anatomy and Physiology 34
Inspection and Palpation 38
Auscultation of the Heart 38
Special Cases 49
The Pulse 51
Practical Points 56

4. Vital Sign #3: Respiration 58
Anatomy and Physiology 58
Atypical Breathing 65
Common Examples of Labored Breathing 68
Practical Points 71

5. Vital Sign #4: Blood Pressure 74
Anatomy and Physiology 74
Blood Pressure Devices 75
Indirect Measurement of Blood Pressure 77
Increased Pulse Pressure 80
Decreased Pulse Pressure 81
High Blood Pressure (Hypertension) 81
Hypertensive Emergencies 81
Secondary Hypertension 83
Low Blood Pressure (Hypotension) 84

Hypovolemic Shock 84
Cardiogenic Shock 89
Septic Shock 90
Neurogenic Shock 91
Anaphylactic Shock 91
Other 92
Special Cases 92
Practical Points 93

6. Vital Sign #5: Level of Consciousness 96
Anatomy and Physiology 96
Management of Altered Level of Consciousness 100
Neurological Examination 102
Physical Examination 103
Causes and Treatments of Coma 105
Practical Points 112

7. Pediatric Vitals 113
The APGAR Score 113
Temperature 114
Heart Rate/Pulse 116
Respiration 116
Blood Pressure 120
Level of Consciousness 123
Practical Points 126

8. Resuscitation 128
Adult Resuscitation 128
Basic Life Support (BLS) 128
Advanced Life Support (ALS) 128
Pediatric Resuscitation 144
Pediatric Basic Life Support 144
Pediatric Advanced Life Support 144
Neonatal Resuscitation 149
Special Resuscitation Cases 151

9. Future and Controversies 154
Body Temperature and Thermometers 154
Heart Rate, Respiration and Blood Pressure 155
Level of Consciousness 155
Trauma Scores 156
Pediatric Vitals 157
Resuscitation 158
Other 159

Appendix 162

Index 164

Preface

This book is written for anyone taking vital signs: doctor, resident, intern, medical student, nurse, practical nurse, nursing assistant, home health practitioner, emergency medical technician (EMT), as well as medical office and nursing home personnel, the fire fighter and in some cases the dental and x-ray technician.

The information is the result of teaching anatomy, physiology, pathophysiology and emergency medicine to residents, medical students, nurses and nursing students for 20 years, as well as working as an emergency physician for an equal amount of time.

Vital signs are an essential part of the physical examination of almost every patient (some crusty practitioners would say every patient). An important responsibility of the health professional is to take them accurately. A second, and frequently neglected, one is to promptly notify someone when an abnormality exists, such as the elderly male who presents with severe chest and back pain and high systolic and diastolic pressures (possibly having an aortic dissection), or the elderly person presenting with abdominal pain and hypotension (possibly having a ruptured abdominal aortic aneurysm).

A question is sometimes posed, "Are the vital signs that important? Aren't other assessments equally as important, such as pain, etc?" The answer is that the original reason for the term is that they were vital, that is—significant abnormalities were life-threatening and must be corrected for survival. This concept has not changed.

This book is not designed for the intensive care setting. Many adequate critical-care textbooks are available for information on invasive monitoring.

Certain aspects of the vital signs, such as use of the tympanic thermometer (an investigative project pursued by the author), the management of pediatric fever and the use of antipyretics, are controversial and are discussed in Chapter 9. The reader will note that a 5th vital sign, Level of Consciousness, is the subject of Chapter 6. Level of consciousness has been assessed by prehospital and hospital personnel for many years and has functioned as a vital sign without an official designation. Other topics such as pulse oximetry are discussed in Chapter 9.

At the end of each chapter is a section on rapid evaluations (Practical Points), with pitfalls and suggestions that should be helpful.

Extensive revisions have been done on BLS, ACLS and PALS algorithms in the year 2000 by an International Educational Conference for Emergency Cardiac Care, consisting of the American Heart Association in collaboration with an International Liaison Committee on Resuscitation (ILCOR). Some, to say the least, are puzzlingly complex. This is also discussed in Chapter 9.

Vitals can be deceptive. In the obese, it is sometimes impossible to hear a heart-beat. In the elderly, sometimes neither a radial nor carotid pulse is palpable. Occasionally, it is difficult to know if a person is breathing, let alone alive. This was illustrated not long ago when a first year resident, having found no pulses or respirations in an old man, called a "code" and began performing cardiopulmonary resuscitation. In a few seconds the elderly gentleman rose up and yelled, "Get off me, you!"

Joseph V. Stewart, M.D.

Acknowledgments

To Alexander Lane for recognizing the importance of the vital signs in our earlier anatomy and physiology teaching days, to Ken Smith for his fine art work, to Pam Bartley for her counsel, and to Sarah Gable and Stephanie Elliott for their research help.

1

History of the Vital Signs

The Thermometer and Temperature

The first primitive thermometer, a glass tube with a column of water displaced in proportion to heat applied, was invented by Heron of Alexandria sometime in the 2nd century AD. About 1595, Galileo reintroduced and modified the device. In a letter to Cardinal Cesarini in Rome in 1638, the Benedictine monk Benedetto Castelli wrote, "I remember having seen more than 35 years ago, an experiment performed by our Senor Galileo. He took a little vase of glass, the size of a small hen's egg, with a neck approximately two palms long, and subtle as a stalk of grain. He warmed the little vase well in the palm of his hands. Then he turned it upside down and placed the mouth of the stalk into a vessel below, filled with some water. When he let the little vase go from the warmth of his hands, the water began immediately to rise in the stalk more than one palm above the water level" (Fig. 1.1).

Inspired by the invention of his friend Galileo, Sanctorius (1561-1636), chair of the Theory of Medicine at the University of Padua, described research on body heat and the thermometer in Commentaries on the first section of the first book of Avicenna: "The instrument was used by Hero for other purposes, but I have applied it to the determination of the warm and cold temperature of the air and of all parts of the body, as well as for testing the heat of persons in a fever". In 1617, the word "thermoscope" appeared in print to describe these primitive devices, and in 1624 the word "thermometer" was coined by Leurechon. The early thermometers, or "air thermoscopes", were glass tubes, open at one end, partially filled with air and set in basins of water.

Around 1654, Ferdinand II of Tuscany, of the Medici family, filled a glass tube with colored alcohol and sealed it by melting the tip. The closed instrument was graduated by degrees marked on the stem. This was the first thermometer independent of atmospheric pressure. Ferdinand and his brother Leopold formed a society in 1657, the Academia del Cimento, consisting of nine members, mostly students of Galileo and a few foreign correspondents, for research and to serve as a sanctuary for scientists. The academy met in Florence at the palace of Leopold, who also presided. Five thermometers were developed by the academy. Wine was used rather than water as an expansion fluid because it is "sooner sensible of the least change of heat and cold, and does not freeze in extreme cold". Florentine thermometers became

Vital Signs and Resuscitation, by Joseph V. Stewart. ©2003 Landes Bioscience.

Fig. 1.1 Galileo's Thermoscope—circa 1595. Reprinted with permission from: Benzinger T. Temperature, Part I: Arts and Concepts. ©1977 Dowden, Hutchinson & Ross.

famous throughout Europe. Church authorities who persecuted Galileo caused the academy to be dissolved after ten years, but Florentine thermometers continued to be manufactured into the 18th century (Fig.1.2).

In 1665, the Irish physicist and chemist Robert Boyle used aniseed oil for fixed points on a thermometer scale. At the same time, Robert Hooke, English physicist, mathematician and inventor, established the freezing-point of water as a fixed point. Hooke filled his thermometers with "the best rectified spirit of wine highly ting'd with the lovely colour of cochineal". Around 1701, Isaac

Fig. 1.2. Florentine Thermometer—circa 1660. Reprinted with permission from: Benzinger T. Temperature, Part I: Arts and Concepts. ©1977 Dowden, Hutchinson & Ross.

Newton tried linseed oil as an expansion fluid. For fixed points in the scale he chose the temperature of melting snow and of the human body, dividing the interval into twelve equal parts.

G.D. Fahrenheit, a German instrument-maker, overcame the expansion problem and increased the sensitivity of the thermometer by creating a fine-bore capillary tube. He designed an alcohol thermometer in 1709 and a mercury one in 1714. Fahrenheit chose as zero the lowest temperature of a freezing mixture of ice and salt and increased the 12 divisions suggested by Newton, probably for convenience, to 96. In Philosophical Transactions of the Royal Society of London in 1724, he states: "Mainly I made two kinds of thermometers, of which one is filled with spirit of wine, the other with quicksilver ... the division of their scales is based on three fixed points ... the first is placed at the lowest part or beginning of the scale, and is attained with a mixture of ice, water, and sal-ammoniac or sea-salt; if the thermometer is placed in this mixture, its fluid descends to a point that is marked zero. This experiment succeeds better in winter than in summer. The second fixed point is obtained if water and ice are mixed together without the above-mentioned salts. If the thermometer is placed in this mixture its fluid takes up the thirty-second degree, which I call the point of the beginning of congelation, for in winter stagnant waters are already covered with a very thin layer of ice when the liquid in the thermometer reaches this degree. The third fixed point is found at the ninety-sixth degree; and the spirit expands to this degree when the thermometer is held in the mouth, or under the armpit, of a living man in good health, for long enough to acquire perfectly the heat of the body ..."

Anders Celsius, a Swedish professor of astronomy, in 1741, accepted the suggestions of Huygens and others to use 0 as the boiling point of water and 100 as the temperature of melting ice. The numbers were reversed by Christin of Lyons and the botanist Linnaeus (Carl von Linne) shortly thereafter, and the centigrade scale was created.

The first important user of thermometry in clinical medicine, and a contemporary of Fahrenheit, was a Dutch physician at the University of Leyden, Herman Boerhaave (1668-1738). Temperatures were taken on all patients. Boerhaave's students de Haen and Van Swieten in Vienna furthered the use of thermometry. Boerhaave in 1731 described "an elegant thermometer made by request by the skilled artist Daniel Gabriel Fahrenheit". According to Boerhaave, Fahrenheit's zero coincided with the greatest natural cold observed in Iceland in the winter of 1709, and this is thought to have been the origin of the lower fixed point in the scale.

The thermometer of the late 1700's was a bone scale wired to a glass tube, about 8 inches long and slightly bent 1 to 2 inches from the bulb since it had to remain in the axilla while being read. The labels read: freezing–32, temperate–48, agreeable–64, very warm–80, blood heat–96, fever heat–above 112.

James Currie of Edinburgh in 1805 created a small mercury thermometer with a moveable scale on the surrounding bone collar adapted from a 6.7 inch instrument invented by John Hunter, a London doctor. Readings were obtained by placing the bulb under the tongue and seemed to be equivalent to those taken in the axilla (Fig. 1.3).

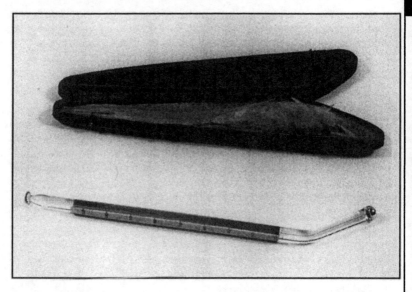

Fig. 1.3. Axillary thermometer and case—circa 1800's. Reprinted with permission from: National Museum of American History, Smitsonian Institution #78-695.

A professor of medicine in Leipzig, Carl Wunderlich, in 1871 published a large treatise, *Das Verhalten der Eigenwarme in Krankheiten (The Behavior of Body Temperature in Disease)*, describing the results of twenty years of experience with the thermometer. "Ever since October 1851 I have introduced the thermometer in my clinic. The number of single observations to some millions, and the number of cases about 25,000." Among other things, Wunderlich

1. established the range of normal body temperature at about 96.8 F (36 C) to 100.4 F (38 C),
2. calculated a mean temperature of 98.6 F (37 C),
3. characterized fever at 100.4 F (38 C) or greater,
4. noted a diurnal variation of normal body temperature (lower in the morning and higher in the afternoon),
5. determined the differences between axillary, oral and rectal temperatures,
6. described the fever patterns of several diseases, and
7. noted that the pulse rises about 9 to 10 beats for every degree Fahrenheit increase in body temperature.

Wunderlich used several types of thermometers and several locations on the body, including the axilla, mouth and rectum. His favorite site was the axilla. In the typical case, the thermometer, which ranged in length from 6 inches to nearly a foot, was left in the axilla for 10 to 20 minutes, depending on the specific thermometer.

Wunderlich comments, "It is then advisable, as *Liebermeister* has recommended, to keep the axilla closed (by bringing the arm to the side) some time before the thermometer is put there. The thermometer is then introduced deep into the axilla (under the anterior or pectoral fold), and the axilla closed, by close pressure of the arm against the thorax. The mercurial column seldom becomes stationary, in measurements taken in the axilla (unless that has been kept closed for some time before) in less than ten minutes, or oftener a quarter of an hour, sometimes it takes twenty minutes, or even longer. The observation may be terminated when the mercury has remained stationary for five minutes." (Fig. 1.4)

William Aitken, professor of pathology at the British army medical school at Chatham and at Netley Hospital, designed and popularized the first self-registering clinical thermometer. Sold in sets that included a straight instrument for oral use and a bent one for the axilla (like a shepherd's crook, according to a nurse at St. Thomas' Hospital), they were about eleven inches long with scales etched on the glass. Aitken made a thermometer for Wunderlich in 1852, as did Thomas Allbutt, a physician at Leeds, who in 1867 developed a short—stemmed thermometer about the size and shape of the glass thermometer in use today.

A German researcher, W.R. Hess, in 1932 and Ranson in 1936 suggested that the area of the brain inferior to the thalamus (the hypothalamus) is responsible for many autonomic functions, including temperature control. The discovery of set-point changes in the hypothalamus by Hardy and Hammell in 1965 led to the observation in 1969 by Eisenman that these changes may account for fevers due to infection. Moore noted in 1970 that toxins called pyrogens are liberated by bacteria and by white blood cells. Vane in 1971 pointed out that a chemical mediary of pyrogens may be prostaglandins.

The glass thermometer has enjoyed widespread use for 50 years. Electronic devices with digital displays and disposable sterile sleeves fitting over oral and rectal probes have been in use since the 1970's. In 1988 it was found that it was possible to measure the amount of infrared radiation emitted from the tympanic cavity, and the first portable tympanic thermometer was invented.

Heart Rate and Pulse

Huang Ti (c. 2600 BC), the last of the Chinese Celestial Emperors, mandated the compilation of an extensive medical treatise, the *Nei Ching* (Yellow Emperor's Book of Medicine). Over 50 pulses and variations were recorded. "Pulses could be sharp as a hook, fine as a hair, dead as a rock, deep

```
A. Collapse temperatures - below 96.8 F (36 C):
     1. Moderate collapse, in itself without danger: 95 - 96.8 F
        (35 - 36 C)
     2. Algide collapse with great danger, still with possibility
        of recovery: 92.3 - 95 F (33.5 - 35 C)
     3. Deep, fatal algide collapse: below 92.13 F (33.5 C)

B. Normal or almost normal temperatures:
     1. Subnormal: 96.8 - 97.7 (36 - 36.5 C)
     2. Really "normal" temperatures: 97.9 - 99.1 F (36.6 - 37.4 C)
     3. Subfebrile temperatures: 99.5 - 100.4 F (37.5 - 38 C)

C. Febrile temperatures:
     1. Slight febrile action: 100.4 - 101.1 F (38 - 38.4 C)
     2. Moderate:
        Morning: 101.3 - 102.2 F (38.5 - 39 C)
        Evening: to 103.1 F (39.5 C)
     3. Considerable:
        Morning: to 103.1 F (39.5 C)
        Evening: to 104 F (40 C)
     4. High:
        Morning: above 103.1 (39.5)
        Evening: above 104.9 (40.5 C)

D. 107.6 F (42 C) and above indicates a fatal termination, except
   in relapsing fever, hyperpyretic temperatures

   -------------------------------------------------------------

E. The temperature in the axilla, in health, under the most varied
   circumstances and influences, moves between 97.2 to 100.4 F
   (36.2 to 38 C)

F. The temperature of the interior or cavity of the mouth, if there
   is no disturbing element, occupies an intermediate position
   between the axillary and vaginal or rectal. We may safely allow
   that when the mean temperature of the axilla in a healthy person
   is 98.6 F (37 C), that of the mouth will be 98.8 to 99 F (37.1
   to 37.2 C), and that of the vagina, or unloaded rectum 99.1 to
   99.5 F (37.3 to 37.5 C)
```

Wunderlich's Classification of Body Temperatures

Fig. 1.4. Wunderlich's Classification of Body Temperatures—1871 (Transl.). Reprinted with permission from: Wunderlich C, Seguin E. Medical Thermometry and Human Temperature. © 1871 William Wood & Co.

as a well, soft as a feather." The volume, strength, weakness, regularity, or irregularity of the pulse revealed the nature of the disease, whether it was chronic or acute, its cause and duration, and the prospects for death or recovery.

An Egyptian papyrus of about the 7th century BC described the pulsations of the heart and the propagation of beats throughout the body: "Its pulsation is in every vessel of every member". Air came in through the nose (but also the ears), entered the channels, was delivered to the heart, and from there was sent to all parts of the body.

The Greeks expanded knowledge of the heart and circulation. Hippocrates (about 460-370 BC) described the pericardium, the ventricles, the heart valves and contracting times of atria and ventricles. Praxagoras of Cos (about 340 BC) separated the functions of arteries and veins, with an emphasis on the pulse. Aristotle (384-322 BC), founder of comparative anatomy, described the early development of the heart and great vessels, the differences between the arteries and veins and named the great arterial vessel the "aorta". Herophilus of Chalcedon (about 280 BC) recognized that the heart transmitted pulsations to the arteries, described the pulse in terms of size, strength, rate and rhythm, and attempted to measure the rate with an improved Alexandrian water clock. Erasistratos (about 250 BC) described the heart chambers and valves.

In the Roman period, Rufus of Ephesus (110-180 AD) reinforced the fact that the heart-beat was the cause of the pulse, and discussed its properties. The Greek physician Galen (129-200 AD), the first experimental physiologist, accurately described the valves of the heart and developed a complicated lexicon of descriptive terms about the pulse.

The astronomers Kepler and Galileo used pendulums and balance clocks to estimate the pulse rate. Galileo timed the swinging chandelier in the Pisa Duoma with his own pulse, counting the rate at eighty beats per minute. When watches with second hands were introduced in the 1690's, physicians could accurately measure the pulse. John Floyer wrote several volumes in 1707 and 1710 on a pulse-timer he called the pulse-watch. By counting the number of pulses per minute, he created a demand for watches capable of registering the time in seconds. In 1768 William Heberden listed pulse rates expected at various ages.

Carl Vierordt in 1828 constructed an instrument designed to trace a graph of the pulse. Called a sphygmograph (Gr: sphygmos = pulse), pulsations were communicated to a lever and the tracings were recorded as vertical strokes (Fig. 1.5).

The French physician Rene Laennec (1781-1826) invented the stethoscope in 1816. Previously, sounds of the lungs and heart were studied by holding one's ear against a patient's chest. "I rolled a quire of paper into a kind of cylinder and applied one end of it to the region of the heart and the other to my ear, and was not a little surprised and pleased, to find that I could thereby perceive the action of the heart in a manner much more clear and distinct than I had ever been able to do by the immediate application of the ear...this first instrument was kept in shape by paste. I now employ a cylinder of wood, an inch and a half in diameter and a foot long, perforated longitudinally by a bore three lines wide, and hollowed out into a funnel-shape, to the depth of an inch and a half at one of its extremities. This instrument I commonly designate simply the Cylinder, sometimes the Stethoscope." (Fig. 1.6)

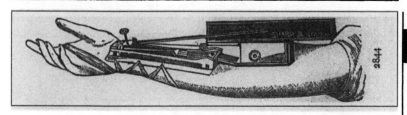

Fig. 1.5. Sphygmograph—1889. Reprinted with permission from: National Museum of American History, Smitsonian Institution #79-5031.

D.J. Corrigan (1802-1880), an Irish clinician, described the characteristic pulse of a disease of the aortic valves.

Adolf Kussmaul (1822-1902), a German physician practicing in Freiburg, known for his work with diabetics (Kussmaul's respirations), wrote three sequential articles in the Berliner Klinische Wochenschrift—Organ fuer practische Aerzte (Berlin Clinical Weekly for Practicing Physicians) in September of 1873 describing the transient disappearance of the pulse during inspiration (paradoxical pulse) in 4 patients with constrictive pericarditis, as well as an inspiratory increase in jugular venous pressure (Kussmaul's sign). "Clinically our affection of chronic inflammation of the pericardium and its obliteration, which is a criterion of mediastinitis, leads to a peculiar pulse phenomenon from time to time associated with unusual behavior of the neck veins. During the time that the sternum with each inspiration exerts a narrowing tug upon the ascending aorta or the arch, the pulse in all the arteries becomes regularly and rhythmically smaller, while the heart movements remain constant. Thus with each inspiration, at regularly repeated intervals, the pulse becomes smaller to return again with expiration. I propose, therefore, to call this the paradoxical pulse, because of the peculiar disproportion between the heart activity and the pulse."

R. Marchand in 1877 recorded potential variations in an exposed frog's heart with a modified galvanometer, or "differential rheotome", obtaining the first electrocardiogram. Augustus Waller, a London physiologist, in 1887 found that the electrical activity of the human heart could be recorded with a modification of the device, or "capillary electrometer". Willem Einthoven of Leiden University in Holland, after working with the capillary electrometer and being dissatisfied with the results, constructed his own galvanometer, a "string galvanometer"—the first electrocardiograph—in 1901. In 1908 the first commercial electrocardiograph was manufactured by Cambridge Scientific Instrument Company of London.

Respiration

Hippocrates in the late 5th century BC observed that the Pneuma of the air is taken in by the lungs. Air along with blood fills the arteries. In a case report in his Corpus at the school at Cos, he described the wife of a friend as

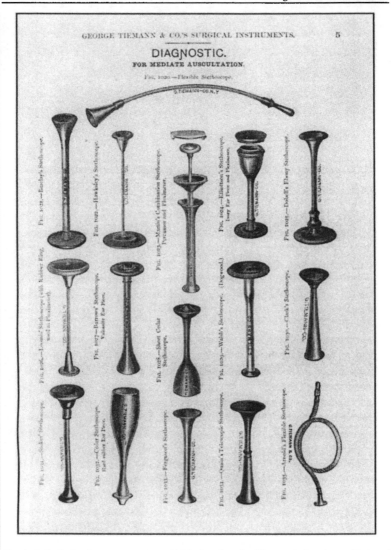

Fig. 1.6. Early Stethoscopes—circa 1889. Reprinted with permission from: National Museum of American History, Smitsonian Institution #80-13427.

having a high temperature with shivering, then, "respiration infrequent, deep for a while and then the breaths would be rapid. She died on the twenty-first day, comatose with deep, intermittent respiration throughout". He was noting a type of periodic breathing seen in the terminally ill and described by John Cheyne (1777-1836), a Scottish physician, and William Stokes (1804-1836), an Irish doctor.

Aristotle (384-322 BC) believed that air is taken into the lungs, absorbed by blood passing through the lungs, and delivered by the pulmonary vein to a fiery heart that the air cools. Herophilus (335-280 BC) claimed that "lungs absorb fresh air and breathe out devitalized air". Erasistratus (310-250 BC) indicated that air is absorbed by the lungs, transported by a "vein-like artery" to the left ventricle to form a "vital spirit", and conveyed by air-filled arteries to various parts of the body. Galen postulated that blood absorbs air into the lungs and is propelled by chest movements through the lungs into a veinlike artery that delivers the mixture to the left ventricle. "There it cools the burning heart, the source of innate heat". Robert Hooke proved in 1667 that air is necessary for life, showing that breathing provides air to the lungs, which converts venous blood into arterial.

Increased interest in the respiratory system took place in the 18th century because of the isolation of oxygen by Karl Scheele (1742-86) and Joseph Priestley (1733-1804). A Scottish chemist, Joseph Black (1728-1799) discovered carbon dioxide. Lavoisier (1743-94) gave the name "oxygen" to the substance in air responsible for combustion and noted that respiration was necessary in living tissue. The French physiologist Claude Bernard (1813-1878) experimented with oxygen and carbon dioxide in the biological system.

Accurate localization of the respiratory center in the medulla was achieved in 1824 by M.J. Flourens and in 1832 by the English physiologist Marshall Hall. The spirometer was invented in 1846 by Hutchinson to measure lung volumes in various groups of people in London, "including 121 sailors, 24 pugilists and wrestlers and 4 giants and dwarfs". Proof that the function of hemoglobin was to take up oxygen was provided by Felix Hoppe-Seyler in 1862.

J.S. Haldane (1860-1936), a Scottish physiologist, discovered respiratory principles as an outgrowth of problems connected with coal mining in England. He found that breathing was regulated by the tension of carbon dioxide in the blood on the respiratory center in the brain rather than by the oxygen, and developed an apparatus for the investigation of respiration and for blood-gas analysis.

Blood Pressure

Galen (mentioned earlier) demonstrated that arteries contain blood, not air, as was thought for 400 years. He grasped the main principles of the venous and arterial circulations, although he incorrectly postulated that the septum of the heart possessed small micropores which allowed blood to move from the right to the left side of the heart.

Nothing much was done after Galen to explore the course of blood flow until the second major experimental physiologist, William Harvey (1578-1657) in 1616 elucidated the mechanism of the pulse and proved that blood circulated within a closed system. He concluded in 1628 in the *Exercitatio*

Anatomica de Moto Cordis et Sanguinis in Animalibus that "the movement of the blood is constantly in a circle, and is brought about by the beat of the heart". Harvey wondered how Galen, having gotten so close to the answer, did not arrive at the concept of a closed circulation.

In 1733, an English clergyman named Stephen Hales, measuring physiological parameters in plants and animals, recorded the first blood pressure. "I endeavoured to find what was the real force of the blood ... in December I laid a common field gate on the ground, with some straw upon it, on which a white mare was cast on her right side ... then laying bare the left carotid artery, I fixed to it towards the heart a brass pipe whose bore was one sixth of an inch in diameter and to that the wind-pipe of a goose; to the other end of which a glass tube was fixed, of nearly the same diameter, which was 12 feet 9 inches long; then untying the ligature on the artery, the blood rose in the tube 9 feet 6 inches perpendicular above the level of the left ventricle of the heart; but it did not attain to its full height at once; it rushed up about half way in an instant, and afterwards gradually at each pulse 12, 8, 6, 4, 2, and sometimes 1 inch; when it was at its full height, it would rise and fall at and after each pulse 2, 3, or 4 inches; and sometimes it would fall 12 or 14 inches, and have there for a time the same vibrations up and down, at and after each pulse, as it had, when it was at its full height; to which it would rise again, after forty or fifty pulses." (Fig. 1.7)

Hales noted that the pulse rate is inversely related, and the blood pressure directly related, to body size. He also found that the cardiac output is the product of the pulse rate and the volume of the left ventricle, and concluded that the blood pressure was the result of the dilation and constriction of blood vessels.

In 1876, Ritter von Basch built an apparatus that indirectly measured the blood pressure of man. This "sphygmomanometer" was the forerunner of a simpler device constructed by an Italian physician, Scipione Riva-Rocci, in 1896. In 1905, Nicolai Korotkoff, a Russian physician, described taking a man's blood pressure, the sequence of which is applicable today: "The sleeve of Riva-Rocci is placed on the middle 1/3 of the arm toward the shoulder. The pressure in the sleeve is raised quickly until it stops the circulation of the blood beyond the sleeve. Thereupon, permitting the mercury manometer to drop, a child's stethoscope is used to listen to the artery directly beyond the sleeve. At first no audible sound is heard at all. As the mercury manometer falls to a certain height the first short tones appear, the appearance of which indicates the passage of part of the pulse wave under the sleeve. Consequently the manometer reading at which the first tones appear corresponds to the maximum pressure. With a further fall of the mercury systolic pressure murmurs are heard which change again to a sound (secondary). Finally, all sounds disappear. The time at which the sounds disappear indicates a free passage of

Fig. 1.7. Hale's first blood pressure recording—1733. Reprinted with permission from: Lyons A, Petrucelli J. Medicine: An Illustrated History. ©1987 Abradale Press/ Abrams, Inc.

the pulse wave; in other words, at the moment the sounds disappear, the minimum blood pressure in the artery exceeds the pressure of the sleeve"

Claude Bernard, a third major physiologist, in 1852 discovered the vaso-constrictor nerves. Etienne Marey, a French physician, in 1859 elucidated the inverse relationship between blood pressure and heart rate. Ernest Starling (1866-1927), an English physician, found that the cardiac output per beat is directly proportional to diastolic filling. Hering in 1927 elucidated the baroreceptor system.

Origin of the Term "Vital Signs"

Because of the invention of the mercury thermometer by Fahrenheit and its gradual modification for clinical use, a renewed interest in body temperature and its relationship to the pulse and respiratory rate came about in the late eighteenth and early nineteenth centuries. In the 7th century, a fast pulse, not an increase in temperature, was the main criterion for diagnosing fever. Interestingly, this notion persisted throughout the 18th century. Boerhaave asserted that a rapid pulse was pathognomonic of fever. Body temperature was a side-issue.

John Hunter (mentioned earlier) in a treatise on the blood, inflammation, and gun-shot wounds in 1794, wrote of a relationship between pulse

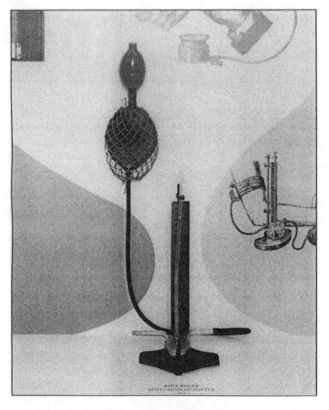

Fig. 1.8. Early Sphygmomanometer—1896. Reprinted with permission from: National Museum of American History, Smitsonian Institution #63367-D.

and temperature in several invertebrate species, as well as dogs, asses, frogs and humans. James Currie (mentioned earlier) in 1797 made simultaneous measurements of the pulse and temperature. In 1818, John Cheyne (mentioned earlier) collected data on the pulse, respiratory rate and temperature on certain patients admitted to the Hardwicke Fever Hospital. Alfred Donne of Paris in 1835 published a series of similar measurements on the relationships between pulse, body temperature, and the respiratory rate.

John Davy in 1863, an English physician, began recording pulse, respirations and temperatures concurrently, noting an "intimate connexion between them". At about the same time, Joseph Jones, professor of medicine at the Medical College of Georgia, presented case reports with observations on the temperature, pulse and respirations in patients with malaria.

Two English translations of Wunderlich's book were published in 1871, one by W.B. Woodman, assistant physician to the London Hospital, and a second by Edouard Seguin, a French psychiatrist and educator who immigrated to America in 1848. Seguin was responsible, not only because of his translation of Wunderlich's book, but through a series of articles and his own book, *Medical Thermometry and Human Temperature,* for introducing the thermometer into usage in the United States. Wunderlich indicated that, "Seguin has made our experience well known in America".

In May, 1866, Edward Seguin, son of Edouard, while an intern at New York Hospital, in collaboration with a colleague William Draper, published an article in the Chicago Medical Journal reporting three cases of pneumonia, including a picture of a chart, "believing that the matter may prove of interest ... the following cases ... are accompanied by a diagram, fac simile of the tables of 'Vital Signs,' used at the bedside to make the daily record of temperature, pulse-beats and respirations." It is labeled: **Record of Vital Signs** (Fig. 1.9).

In October of the same year, a report in the New York Medical Journal by Austin Flint, professor of medicine at Bellevue Hospital Medical College, described a similar chart developed by J.M. DaCosta, physician to the Pennsylvania Hospital. Vital sign recording became a part of clinical practice at Bellevue and New York hospitals in 1867.

Sometime in the early 20th century, blood pressure became a vital sign. The phenomenon occurred subtly, almost imperceptibly, and its inclusion seemed to depend on the type of hospital and patient acuity. In charts of the 1950's, sometimes blood pressure was included as a vital sign, sometimes not. In almost all charts of the 1970's, blood pressure was included as a vital sign (although, interestingly, in many hospitals today it is not included with the vital signs). Level-1 trauma hospitals used it, a phenomenon not unlike what is happening with level of consciousness today.

Level of Consciousness

An interest in coma (Gr: deep sleep) began in the 1960's. In 1966, A.K. Ommaya in England described a five-point level of consciousness scale in a

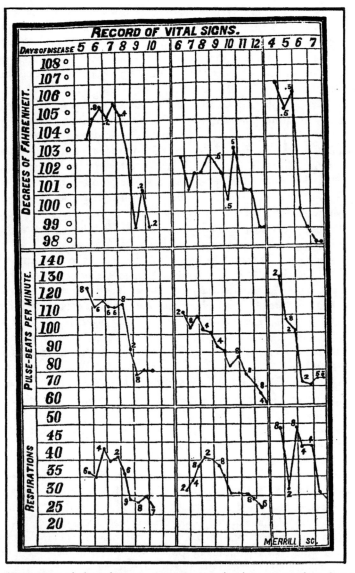

Fig. 1.9. First Record of "Vital Signs"—1866. Reprinted with permission from: Seguin E. The use of the thermometer in clinical medicine. Chic Med J 1866; 23:193.

study of head trauma. In 1968, a neurosurgical "watch sheet", consisting of 7 parameters for evaluating clinical improvement/deterioration of brain-injured patients: ability to move, pupillary reaction, nonverbal reaction to pain, ability to awaken, speech, consciousness and vital signs, was published in the Journal of Trauma by W.F. Bouzarth, a Pennsylvania neurosurgeon. Also in 1968, a digital scale evaluating 9 levels of response was developed by the Birmingham Accident Hospital in Leeds, England (see Fig. 9.1).

A significant development, in 1970, was a series of discussions between Fred Plum, chief neurologist at New York Hospital and two neurosurgeons from Glasgow University, Graham Teasdale and Bryan Jennett, about whether various treatments made a difference in the outcomes of coma patients. A multinational investigation was undertaken to study the progress of comatose patients.

After reviewing 14 "responsiveness" or "coma" scales, Jennett and Teasdale published "A Practical Scale" in The Lancet in 1974 analyzing three aspects of behavior: motor response, verbal response and eye opening. Originally graded from 3 to 14, a 'best motor response' later increased the maximum score to 15, where it is now. This Glasgow Coma Scale (GCS) has served a useful purpose for nearly three decades as a model for the evaluation of level of consciousness not only from trauma, but from metabolic, vascular and infectious causes as well (Fig 6.2).

Although the GCS was originally designed to evaluate deterioration and improvement in coma, it has now become widely used for evaluation and management of level of consciousness by EMTs and emergency personnel. A Pediatric Glasgow Coma Scale has materialized. The GCS is an integral part of the Revised Trauma Score and is a component of the CRAMS Scale (see Figs. 6.3, 7.8, 7.9 and 9.3).

References

1. Anning S. Clifford Allbutt and the clinical thermometer. The Practitioner 1966; 197:818.
2. Benzinger T. Temperature, Part I: Arts and Concepts. Stroudsburg: Dowden, Hutchinson & Ross, 1977.
3. Bloch H. Phenomena of respiration: Historical overview to the twentieth century. Heart Lung 1987; 16:419.
4. Bolton H. Evolution of the Thermometer. Easton: The Chemical Publishing Co., 1900:1592-1743.
5. Bouzarth W. Neurosurgical watch sheet for craniocerebral trauma. J Trauma 1968; 8:29.
6. Burch G, DePasquale N. A History of Electrocardiography. Chicago: Year Book Medical Publishers, 1964.
7. Clark-Kennedy A. Stephen Hales—An Eighteenth Century Biography. Cambridge: At the University Press, 1929.
8. Comroe J. The soul of wit. In: Retrospectroscope: Insights into Medical Discovery. Menlo Park: Von Gehr Press, 1977.
9. Cournand A. Historical details of Claude Bernard. Trans NY Acad Sci 1, 1980.
10. Davis A. Medicine and its Technology. Westport: Greenwood Press, 1981.

11. Dominguez E et al. Adoption of thermometry into clinical practice in the United States. Rev Inf Dis 1987; 9:1193.

12. Dominguez E, Musher D. Clinical thermometry. In: Mackowiak P, ed. Fever. New York: Raven Press, 1991.

13. Estes J. Quantitative observations of fever and its treatment before the advent of short clinical thermometers. Medical History 1991; 35:189.

14. Fishman A, Richards D. Circulation of the Blood, Men and Ideas. New York: Oxford Univiversity Press, 1964.

15. Gauchat H, Katz L. Observations on pulsus paradoxus. Arch Intern Med 1924; 33:350.

16. Hering H. Die Karotidssinus Reflexe auf Herz und Gefaesse. Leipzig: D. Steinkopff, 1927.

17. Hess W. Hypothalamus and Thalamus, Experimental Documentation. Stuttgart: George Thieme, 1956.

18. Heymans C, Neil E. Reflexogenic Areas of the Cardiovascular System. London: Churchill, 1958.

19. Hutchinson J. On the capacity of the lungs. Med Chir Trans 1846; 29:137.

20. Kussmaul A. Ueber schwielige Mediastino—Pericarditis und den Paradoxen Puls. Berl Klin Wochenschr 1873; 10:37-39.

21. Kussmaul A. Jugenderinnerungen (Memoirs of an Old Physician). Stuttgart: Adolf Bons and Co., 1899./Translated by W. Stewart/ The Israel program for scientific translations, Jerusalem; New Delhi: Amerind Publishing Co., 1981.

22. Laennec R. Treatise on the Diseases of the Chest and on Mediate Auscultation. New York: Samuel Wood & Sons, 1830.

23. Lyons A, Petrucelli J. Medicine: An Illustrated History. New York: Abradale Press/ Abrams, Inc., 1987.

24. Mackowiak P, Worden G, Carl Reinhold. August Wunderlich and the evolution of clinical thermometry. Clin Inf Dis 1993; 18:458.

25. Magner L. A History of Medicine. New York: M Dekker, Inc., 1992.

26. Margotta R. The Story of Medicine. New York: Golden Press, 1968.

27. Middleton W. A History of the Thermometer. Baltimore: The Johns Hopkins Press, 1966.

28. Mitchell S. The Early History of Instrumental Precision in Medicine. New York: Lenox Hill Pub., 1892.

29. Musher D, Dominguez EA, Bar-Sela A. Edouard Seguin and the social power of thermometry. N Engl J Med 1987; 316:115.

30. Ommaya A. Trauma to the nervous system. Ann Roy Coll Surg Engl 1966; 39:317.

31. Price D. Factors restricting the use of coma scales. Acta Neurochir Suppl 1986; 36:106.

32. Ranson S. Some functions of the hypothalamus. Bull NY Acad Med 1937; 13:241.

33. Seguin E. The use of the thermometer in clinical medicine. Chic Med J 1866; 23:193.

34. Seguin E. Opera Minora. New York: GP Putnam's Sons, 1884.

35. Stein D. Historical perspective on fever and thermometry. Clin Ped (suppl) 1991.

36. Sternbach G. The Glasgow Coma Scale. J Em Med 2000; 19:67.

37. Teasdale G, Jennett B. Assessment of coma and impaired consciousness: A practical scale. Lancet 1974; 2:81.

38. Teasdale G, Jennett B. Assessment and prognosis of coma after head injury. Acta Neurochir 1976; 34:45.

39. West J. Best and Taylor's Physiological Basis of Medical Practice. Baltimore: Williams & Wilkins, 1991.

40. Wilbur C. Antique Medical Instruments. Atglen: Schiffer Pub., 1987.

41. Wilks S. The scale of Fahrenheit's thermometer. Brit Med J 1900; 2:1212.
42. Wunderlich C. Das Verhalten der Eigenwarme in Krankheiten (The Behavior of Body Temperature in Disease). London: New Sydenham Society, 1871.
43. Wunderlich C, Seguin E. Medical Thermometry and Human Temperature. New York: William Wood & Co., 1871.

1

CHAPTER 2

Vital Sign #1: Temperature

2

Man, other mammals and birds are "warm-blooded" (homeothermic—Gr: unchanging temperature), meaning that although exposed to a variety of temperatures the body remains at a certain warm temperature. Cells of the body function optimally within a narrow range (97.7-100.4°F/ 36.5-38°C), and mammals and birds have evolved a way of regulating the internal temperature to stay within this span. In contrast, amphibians, reptiles and fish ("cold-blooded" or poikilothermic—Gr: variable temperature) possess less sophisticated regulatory systems, take on the temperature of the environment and must take radical moves to avoid extremes of hot and cold (actually "cold-blooded" is a misnomer, since frogs, snakes and fish have warm blood).

The temperature regulatory center in man is the **hypothalamus**, an area of the brain above the pituitary gland which acts as a thermostat to maintain the body's internal, or core, temperature at a range of about 97-99.8°F/ 36.1-37.7°C (Figs. 2.1, 2.2).

Heat Production and Loss

The body strives to maintain a balance between heat production and loss. Heat is produced by muscular activity and cell metabolism. It is lost by radiation, evaporation, conduction and convection. Radiation, the loss of heat in the form of infrared waves, accounts for 60% of body heat loss. Sweating and evaporation account for 25% of body heat loss. Conduction, the transfer of heat from one object to another of lower temperature, and convection, the removal of heat by air currents, are not significant sources of heat loss.

Perception of and Reaction to Hot and Cold

Perception of Hot and Cold

Perception of hot and cold is by way of warm and cold receptor neurons in the hypothalamus. A change in the temperature of blood flowing through the hypothalamus activates mechanisms to stabilize body temperature. In addition, warm and cold receptors in the skin and elsewhere send messages to the hypothalamus to modify the temperature. Temperature and pain fibers of the lateral spinothalamic tract run to the thalamus, and from there via the

Vital Signs and Resuscitation, by Joseph V. Stewart. ©2003 Landes Bioscience.

Fig. 2.1. Centigrade/Fahrenheit scales.

Centigrade	Fahrenheit
43°	109.4°
42	107.6
40.5	105.8
40	104.9
39.5	103.1
39	102.2
38.5	101.3
38	100.4
37.5	99.5
37	98.6
36.5	97.7
36	96.8
35.5	95.9
35	95
34	93.2
33	91.4
32	89.6
31	87.8
30	86
25	77
20	68
15	59
10	50
+5	41
0	32

$1°F = 0.54°C \qquad 1°C = 1.8°F$

To convert fahrenheit into centigrade, subtract 32 and multiply by 0.555.

To convert centigrade into fahrenheit, multiply by 1.8 and add 32.

thalamic radiation to the postcentral gyrus of the parietal lobe of the cerebral cortex. Connections exist between the postcentral gyrus, the thalamus, the reticular formation and the hypothalamus.

Reaction to Cold

Blood vessels are innervated by sympathetic fibers of the autonomic nervous system, with hypothalamic connections. When it is cold, sympathetic fibers are stimulated and vessels constrict (vasoconstriction), conserving heat.

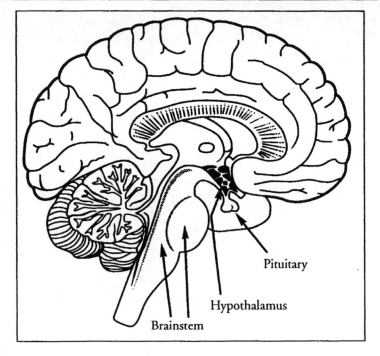

Fig. 2.2. Hypothalamus.

The person becomes more active (i.e., pacing, arm movements). Signals are relayed down the spinal cord via the lateral columns and muscle tone is increased. As the tone rises, involuntary contraction and twitching occur (shivering). Shivering may increase body-heat to five times normal.

Reaction to Heat

When it is hot, sympathetic stimulation to vessels is inhibited and vessels widen (vasodilation). Core body heat as sweat, an ultrafiltrate of blood, moves through dilated vessels and sweat glands to skin, where evaporation takes place. Body activity decreases to prevent unnecessary heat production. When an adult is not sweating, about 700 ml of body water is lost from the skin and lungs by evaporation each day (insensible water loss). On a hot day, an active person may lose over a liter of fluid per hour by sweating. In a humid environment, poor evaporation causes body heat to remain high, contributing to heat illness.

Acclimatization

Acclimatization is the adaptation of the body to continuous heat or cold over several weeks. In man, heat adaptation is more developed than cold adaptation. Aerobic capacity increases. In a hot environment, heat production decreases and heat loss increases for a given amount of work. Sweat production may increase to 2 liters per hour. Increased aldosterone secretion causes decreased salt loss. In a cold environment, heat production may be increased by 10%.

Body Temperature

The concept of a "normal" body temperature is frequently reevaluated. In spite of new instruments and studies, the results remain fairly constant. Minor differences exist because of genetics, sex, age and body size. No two studies agree on the range of normal oral, rectal and core temperatures. However, as Wunderlich implied, some increases and decreases should probably be included in the normal physiological range. Current evidence indicates that:

1. The normal **core** temperature range, verified by indwelling catheters in the esophagus, urinary bladder, pulmonary artery and tympanic cavity, is from about 97.0-$99.8°F$ (36.1—$37.7°C$). Slightly higher or lower normal temperatures exist.

2. **Oral** and **axillary** temperatures parallel the core range. Occasionally the readings are 1-$2°F$ (0.5-$1.1°C$) lower than core. Low readings may reflect user error. A recent study has suggested that the mean oral temperature is $98.2°F$ ($36.8°C$), not $98.6°F$ ($37°C$). Although axillary temperatures were the first used clinically, they have become unreliable and must be verified by another method (see next section, and Chapter 9).

3. **Rectal** temperatures parallel the core range. Readings are sometimes erratic, and are occasionally 1-$2°F$ (0.5-$1.1°C$) higher than core.

4. A 0.1-$2.4°F$ (0.05-$1.3°C$) diurnal range is present, with a low at 6 AM and a high at 6 PM.

5. Temperatures may drop to $97°F$ ($36.1°C$) during sleep and in cold weather. Hot weather or strenuous activity may raise the temperature to 101-$102°F$ (38.3-$38.9°C$).

6. Pediatric temperatures fluctuate, but generally parallel adult readings.

Methods of Temperature Measurement

The Mercury-glass thermometer is the standard to which other methods of thermometry are compared. Today in most clinical circumstances electronic thermometers with digital displays and disposable sleeves are used. A beep or the cessation of blinking indicates the reading. The device does not measure tissue temperature, as does the glass thermometer, but rather the temperature of the probe after contact with target tissue. It does not wait for an equilibrium between the probe and target tissue; rather, it predicts the equilibrium temperature with algorithmic calculations. Although subject to error, it has proved reliable.

Oral

The probe should be under the tongue touching the sublingual gland with the mouth tightly closed for at least 4 minutes with the glass thermometer and an appropriate amount of time for the electronic probe. Factors affecting an oral temperature include rapid breathing, the ingestion of hot or cold beverages, and incorrect placement of the thermometer tip.

Axillary

The axillary temperature requires a closed axilla for 10 minutes before placement of the thermometer or electronic probe, and four minutes is required after placement. Because of the potential for error this is usually impractical. It is used in the trauma patient when other sites are unavailable, and occasionally in pediatrics (see Chapter 7).

Rectal

The rectal temperature is often avoided because of patient discomfort and the extra effort required. Minimal problems occur if the probe is well lubricated. The rectal temperature is least influenced by error. The patient is positioned on the left side with the right thigh and knee flexed. Small children and infants are positioned on their backs and the thighs and knees are flexed while they are held. Children can sometimes be held on their hands and knees. The probe or thermometer is inserted 1-1/2 inches for adults, 1 inch for children, 1/2 inch for infants, and the thermometer is read in 3 minutes. Contraindications are hemorrhoids and rectal disease. Rectal temperatures may be hazardous in newborns because of the possibility of perforation.

Vaginal

The vaginal temperature, similar to the rectal range, is used when other sites are unavailable, as in trauma. It is contraindicated in the last trimester of pregnancy and in uterine and vaginal disease.

Tympanic

The tympanic thermometer assesses infrared emissions from the tympanic cavity. A heat-sensitive probe, covered by a disposable plastic speculum or a plastic-wrap type of material, is inserted into the auditory canal, similar to how one uses an otoscope. The scan button is pressed, the signal is digitized through a computerized processor, and in 2 seconds the temperature is displayed on a small liquid-crystal screen. The probe must make a tight seal in the auditory canal or the reading will be in error. It is useful in the pediatric population, and in situations such as a wired jaw, the multiple trauma patient, the obtunded elderly and the comatose patient. The only contraindication is extensive bilateral auricular malformation/disease. Otitis externa or media are not contraindications. Controversies regarding use and evaluation of the instrument are discussed in Chapter 9 (Fig. 2.3).

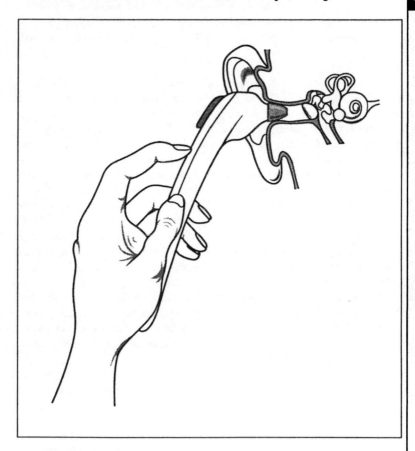

Fig. 2.3. Tympanic Thermometer.

Other

Other areas are used from time to time, such as forehead temperatures in children. Plastic strips had been used, but were considered unreliable. Recently a forehead temporal artery device was evaluated but was also considered clinically unacceptable.

High Temperature

Hyperthermia occurs when heat production exceeds heat loss. It is a failure of the body's heat dissipating mechanisms and is often seen in the unacclimatized individual, particularly the elderly. Occasionally an adverse genetic predisposition to certain drugs will cause hyperthermia. **Fever**, on the other hand, does not involve exposure to heat but rather is the reaction of the body to infection (occasionally other causes). It rarely involves failure of the heat dissipating mechanisms unless the temperature is high and prolonged.

Fever

Fever is present when the core temperature is about 100°F (37.8°C) (99°F/ 37.2°C in AM, 100°F/37.8°C in PM). The hypothalamus maintains the internal or core temperature at a set point of about 97-99.8°F (36.1-37.7°C). Fever resets the set point to a higher degree. Thus, even though the body is hot, it feels cold because the set-point has been raised. Mechanisms to conserve heat, such as vasoconstriction and shivering, are activated (Fig. 2.4).

Bacteria, viruses, phagocytized breakdown products and toxins stimulate monocytes and macrophages to release endogenous pyrogenic cytokines. Antigen-antibody complexes, metabolites and dead tissue sometimes evoke the same response. The main pyrogenic cytokines are interleukin-1, interleukin-2 and tumor necrosis factor. These cytokines induce formation of a prostaglandin (PGE-2) in the temperature center (thermoregulatory center) of the hypothalamus. Prostaglandin resets the normal set point (~98.6°F) to a higher degree.

The majority of fevers are caused by infection, but some are the result of malignancy, stroke and an assortment of allergic and autoimmune disorders. In the typical case, microorganisms are killed, antibody production, an important facet of the immune response, is enhanced and white blood cells show increased phagocytic activity.

Since fever activates mechanisms that increase metabolism and heat production, there is an increase in heart rate, respiratory rate and blood pressure (primarily systolic). The heart rate is elevated by about 9-10 beats for each degree Fahrenheit increase in temperature. The mental status is sometimes affected but brain damage does not occur unless symptoms of heat stroke appear (sustained temperature over 106°F/41.1°C), or meningo-encephalitis is also present.

Antifever drugs (antipyretics) such as aspirin, acetaminophen (Tylenol) and nonsteroidal anti-inflammatory drugs such as ibuprofen (Advil, Motrin) and naproxen (Naprosyn, Anaprox) act on the hypothalamus by blocking prostaglandin synthesis. Steroids block both prostaglandins and endogenous pyrogen, but are weak antipyretics.

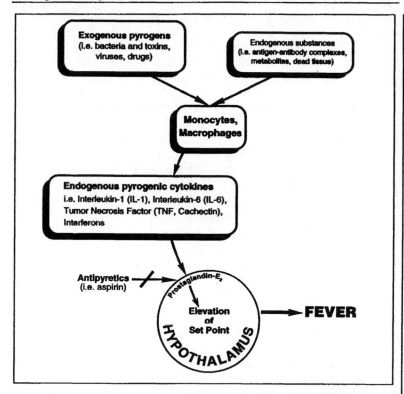

Fig. 2.4. Mechanisms of Fever.

Antipyretics are routinely administered for fever, although the increase in temperature is a primary defense against infection. The body raises the set-point to kill microorganisms which survive poorly in a hot environment. As the temperature increases, phagocytic activity and antibody production increase. When the temperature falls after the administration of antipyretics, phagocytic activity and antibody production decrease. Recent studies have shown that the administration of antipyretic agents prolong the course of infection. It is reasonable to use antipyretics to treat uncomfortable patients and those with temperatures approaching hyperthermia (105°F, 40.6°C) (use of antipyretics in pediatrics is discussed in Chapter 7, and the controversy is further examined in Chapter 9).

Heat Stroke

Heat stroke is a failure of the body to rid itself of heat. In this condition, heat-loss mechanisms fail in response to increased environmental temperature. It differs from fever in that there is no release of endogenous pyrogen or resetting of the hypothalamic set-point. It is seen in those whose heat dissipating mechanisms are compromised, such as infants and the elderly, the obese, alcoholics, those taking certain drugs (i.e., phenothiazines, anticholinergics, antihistamines, sympathomimetics) and in some engaged in heavy exertion (i.e. marathon runners). Anticholinergics, diuretics, phenothiazines, and antihistamines suppress the sweating process (Fig. 2.5).

1. **Drugs =**
   a. **Inhibit sweating:**
      anticholinergics (i.e. atropine)
      antihistamines
      phenothiazines
      tricyclic antidepressants
   b. **Stimulants/hallucinogens:**
      cocaine, LSD, PCP
   c. **Alcohol**
   d. **Diuretics**
2. **Drug withdrawal**
3. **Cardiovascular disease**
4. **Fatigue**
5. **Excessive clothing**
6. **Extremes of age**
7. **Obesity**
8. **Dehydration**
9. **Exercise**
10. **Diseases:**
    a. **Scleroderma**
    b. **Cystic fibrosis**
11. **Debilitated Patients**

Fig. 2.5. Risk Factors for Hyperthermia.

When the temperature approaches $106°F$ ($41°C$), tachycardia and weakness occur. Neurological changes appear, ranging from disorientation and bizarre behavior to seizures and unconsciousness. Although sweating may occur initially, the sweating process eventually fails and the skin is hot and dry. Cells are damaged, proteins are denatured, mitochondria and cell membranes are destroyed and hemorrhages occur. Complications include shock, brain damage, acidosis, muscle cell disruption, kidney and liver failure and intravascular coagulation.

The severity of the outcome is a function of the age and health of the patient, medicines taken and degree of acclimatization. Early death is from cerebral edema, brain cell damage and circulatory failure. Later problems involve the heart, central nervous system and kidneys from rhabdomyolysis and acute tubular necrosis. **Treatment:** in the field, immediate cooling is mandatory. The person is moved to a cool environment, the clothing is removed and he is splashed or sprayed with normal-temperature water. The ABCs are followed (see Fig. 8.15). In the emergency department normal saline is administered at 1 liter per hour. Fanning is begun and ice-packs are applied to the groin and axillae. Treatment is discontinued when the core temperature is 100.4°F (38°C).

Heat Exhaustion

Heat exhaustion is volume depletion, which may lead to shock. The person loses salt and water in various combinations without adequate replacement. It occurs in a hot environment over a longer period of time than heatstroke, in the unacclimatized individual engaged in strenuous physical activity, and in the elderly. The temperature is normal or slightly elevated. In contrast to heat stroke, sweating is present, and the person is often cool and clammy. Neurological symptoms are absent. It may progress to heat stroke. **Treatment:** 4 liters of IV normal saline or Ringers lactate administered over 3 hours.

Heat Cramps

Heat cramps are caused by salt depletion from sweating. It is seen after strenuous physical exercise and involves painful spasms of leg muscles from hyponatremia which interferes with calcium-dependent muscle relaxation. The temperature is normal. **Treatment:** 2 liter bolus of IV normal saline.

Uncommon Heat Illnesses

Malignant Hyperthermia

This rare condition is seen sometimes in a patient undergoing general anesthesia. Muscles become rigid and the temperature rises, sometimes to 107.6°F (42°C). The mechanism appears to be a genetic muscle defect permitting the inappropriate release of calcium from cells. **Treatment:** general anesthesia is stopped, cooling is undertaken as for heat stroke, and a muscle relaxant (dantrolene sodium) is given as a 2 mg/kg IV bolus.

Neuroleptic Malignant Syndrome

A similar rare situation exists with some patients on phenothiazines. The temperature rises, the muscles become rigid, and autonomic instability (tremors, tachycardia, sweating) and confusion appear. The mechanism seems to be dopamine receptor blockade producing muscle spasticity and heat production (up to 104°F/40°C). **Treatment:** cooling is begun as for heat stroke, IV normal saline is given and dantrolene is administered as a 1 mg/kg IV bolus.

Low Temperature (Hypothermia)

Besides exposure to cold, including submersion, other causes or risk factors for hypothermia are sepsis, particularly in the elderly, endocrine insufficiencies such as hypothyroidism, hypoadrenalisms, hypopituitarism, hypoglycemia, as well as ethanol, sedative-hypnotics, opioids and drugs of abuse, blunting the awareness of temperature.

Hypothermia is a core temperature of 95°F (35°C) or less (some use 96°F/35.6°C or less). As the body temperature declines, efforts to increase the temperature are activated: shivering, increased activity and vasoconstriction occur. As the temperature falls, these compensatory measures fail. At 90°F (32.2°C) metabolism slows and the mental status/level of consciousness is affected. Shivering ceases at 86°F (30°C). Below this, cardiac arrhythmias may occur. At 80°F (26.7°C) respiratory and heart rates slow, blood pressure falls and consciousness is lost. Successful rewarming may occur even with a core temperature of 75°F (24°C).

The thermometer must read low. Most glass thermometers read to 94°F (34.6°C). Electronic thermometers, on the other hand, read to 84°F (28.9°C). A rectal or tympanic temperature is necessary for a core reading. A low-reading rectal probe is part of many rewarming mattress devices (i.e., K-thermia). The machine records the rectal temperature, and the rewarmer is set to the required temperature. **Treatment** depends on the degree of hypothermia. Mild cases (90-95°F/32-35°C) respond to rewarming with rewarming blankets. More severe cases (less than 86°F/30°C) may require combinations of the ABCs of resuscitation (Fig. 8.15), warm IV fluids heated to 104°F (40°C), heated humidified oxygen, and peritoneal, gastric and bladder lavage (also heated to 104°F (40°C).

Pulseless nonbreathing hypothermic patients should be resuscitated while rewarmed. Even though a person appears lifeless, rewarming may result in complete recovery. Recovery has been documented in cases of hypothermic cardiac arrest for 3 hours and in cold water submersion for 40 minutes. Resuscitative measures should be undertaken until the core temperature is 90°F. Death in hypothermia is failure to revive after rewarming (Fig. 2.6).

Certain aspects of cold injury (local hypothermia) may accompany hypothermia. **Chilblain** is exposure of a body part (nose, fingers, toes) to above-freezing cold, causing itchy lesions. **Treatment** is warming at room temperature. **Frostbite**, on the other hand, is tissue damage from freezing cold. With a severe wind-chill, even exposure for a few minutes may produce white insensitive areas of skin. The degree of frostbite is comparable to the degree of a burn (first degree, second degree, etc.). **Treatment** is rewarming of the body part in hot water (about 104°F/40°C) for about 20 minutes.

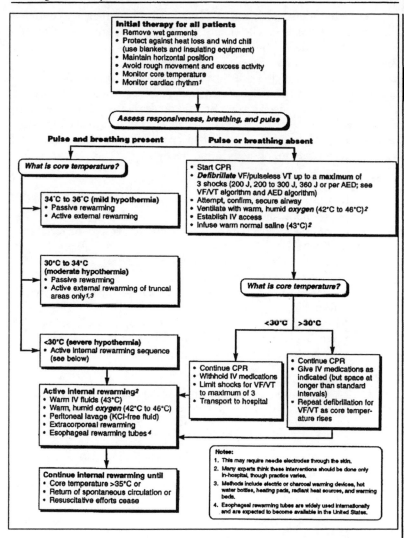

Fig. 2.6. Hypothermia Algorithm. Reprinted with permission from: Guidelines for 2000 for Cardiopulmonary Resuscitation and Emergency Cardiovascular Care, American Heart Association.

Infants and the Elderly

Infants, particularly newborns, and the elderly are prone to hypothermia. In the case of infants the reason is because of the developing hypothalamic thermoregulatory mechanism (see Chapter 7). With the elderly the regulatory mechanism is weakened from aging. Old age causes a diminished ability to perceive and adjust to hot and cold (diminished hypothalamic control of the sympathetic system). The elderly are at high risk for developing life-threatening sequelae of both conditions. This is particularly evident in response to cold. Shivering creates an increase in oxygen consumption and cardiac output, placing the elderly at risk of myocardial infarction, angina and heart and respiratory failure. Dehydration, common in the elderly, raises the temperature, and increases the potential for cardiovascular collapse. Sepsis in the elderly may present with a high temperature, a normal temperature or a low temperature. Hypothermia, hyperventilation and hypotension are common manifestations of sepsis in the elderly.

Practical Points

- Record a temperature on all patients. It is an easy vital sign to forget, for numerous reasons.
- Record the temperature as oral, tympanic, axillary or rectal (i.e., 95°R, 104°O).
- Do not rely on oral temperatures for ill patients. Do a rectal. The patient is often mouth breathing.
- Do not rely on a tympanic temperature in an infant or critically ill patient. Take a rectal.
- Address high (>104°F) and low (<94°F) temperatures without delay. Severe hypo- and hyperthermia need to be treated immediately. Have access to a low-reading thermometer.

References

1. American Heart Association and the International Liaison Committee on Resuscitation (ILCOR): Guidelines 2000 for cardiopulmonary resuscitation and emergency cardiovascular care. Baltimore: Lippincott, Williams & Wilkins, 2000.
2. Ballester J, Harchelroad F. Hyperthermia: How to recognize and prevent heat-related illnesses. Geriatrics 1999; 54:20.
3. Bessen H. Hypothermia. In: Tintinalli J, ed. Emergency Medicine: A Comprehensive Study Guide. New York: McGraw-Hill, 2000.
4. Brady W et al. Life-threatening syndromes presenting with altered mentation and muscular rigidity, Part I: Neuroleptic malignant syndrome, hyperthermia, thyrotoxicosis and malignant catatonia. Em Med Rep 1999; 20:51.
5. Coceani F, Akarsu E. Prostaglandin E-2 in the pathogenesis of fever. Ann NY Acad Sci 1998; 856:76.
6. Danszl D, Pozos R. Accidental hypothermia. N Eng J Med 1994; 331:126.
7. Doyle F. The effect of ambient temperature extremes on tympanic and oral temperatures. Am J Emerg Med 1992; 10:285.
8. Gilbert M et al. Resuscitation from accidental hypothermia of 13.7°C with circulatory arrest. Lancet 2000; 355:375.
9. Kanzenbach T, Dexter W. Cold injuries. Postgrad Med 1999; 105:2.
10. Kluger M et al. Role of fever in disease. Ann NY Acad Sci 1998; 856:224.
11. Lazar H. The treatment of hypothermia. N Engl J Med 1997; 337:1545.
12. Lewit E et al. An evaluation of a plastic strip thermometer. JAMA 1982; 247:321.

13. Lily J et al. Urinary bladder temperature monitoring: A new index of body core temperatures. Crit Care Med 1980; 8:12.

14. Luhishi G. Cytokines and Fever. Mechanisms and sites of action. Ann NY Acad Sci 1998; 856:83.

15. Mackowiak P, ed. Fever: Basic Mechanisms and Management. Philadelphia: Lippincott-Raven Pub., 1997.

16. McGee Z, Gorby G. The diagnostic value of fever patterns. Hosp Pract Oct 30 1987:103.

17. Nierman D. Core temperature measurement in the intensive care unit. Crit Care Med 1991; 19:818.

18. Pidwell W et al. Accuracy of the temporal artery thermometer. Ann Em Med Suppl 2000; 36:5.

19. Saper C, Breder C. The neurologic basis of fever. N Engl J Med 1994; 330:26.

20. Shinozaki T et al. Infrared tympanic thermometer: Evaluation of a new clinical thermometer. Crit Care Med 1988; 16:148.

21. Simon H. Hyperthermia and heatstroke. Hosp Pract 1994; 29:8.

22. Stewart J, Webster D. Re-evaluation of the tympanic thermometer in the emergency department. Ann Em Med 1992; 21:158.

23. Terndrup T. An appraisal of temperature assessment by infrared emission detection tympanic thermometry. Ann Emerg Med 1992; 21:12.

24. Van der Meer J et al. Proinflammatory cytokines and treatment of disease. Ann NY Acad Sci 1998; 856:243.

25. Walker J, Barnes S. Heat emergencies. In: Tintinalli J, ed. Emergency Medicine: A Comprehensive Study Guide. New York: McGraw-Hill, 2000.

26. Willis J, Ji H. Explosive increase in Na+ entry to acidified cells at elevated temperature. Evidence for the energy depletion model of heat stroke? Ann NY Acad Sci 1998; 856:304.

CHAPTER 3

Vital Sign #2: Heart Rate/Pulse

The term "pulse" is a carryover from earlier times, and was used to evaluate the heart-rate. Today both heart-rate and pulse-rate should be assessed, since they may be different. The heart is auscultated for rate, rhythm and extra sounds, such as murmurs. Counting both heart-rate and pulse-rate is easily accomplished at the same time.

The Heart: Anatomy and Physiology

The Heart Rate

The heart rate is the number of double-sounds auscultated for one minute. The first part of the double-sound (1st heart sound, S-1) is the rebound of blood against the heart wall after contraction of the ventricles (systole) and closure of the atrioventricular valves (AV valves—mitral and tricuspid). The second part of the double-sound (2nd heart sound, S-2) is the back-recoil of blood against the closed semilunar valves—pulmonary and aortic—so-called because they are half-moon shaped). The two sounds are magnified by the stethoscope as "lub-dup". In the adult, the average heart rate is about 70 beats per minute. The range is 60 to 100, with exceptions. Below 60 is bradycardia; above 100 is tachycardia. The heart rate or pulse is measured by counting the number of beats for 15 seconds and multiplying by four. If an arrhythmia is suspected, the number of beats is counted for one minute (Fig. 3.1).

Electrical Activity of the Heart

Nerve and muscle cells are specialized for the conduction of electrochemical impulses down the length of the cell. At rest, there is an abundance of sodium on the outside of the cell and an abundance of potassium on the inside. When the cell is stimulated, the impulse proceeds down the cell in a fuse-like fashion. The cell becomes permeable to sodium and sodium flows into the cell. This is depolarization. Potassium then flows out of the cell, restoring electrochemical balance. This is repolarization (later the sodium-potassium pump restores the proper ions to the correct sides of the membrane). During repolarization calcium ions enter the cell by way of channels called "slow channels", or "calcium channels". The conduction system of the heart is activated and calcium initiates contraction of the heart. This electrical

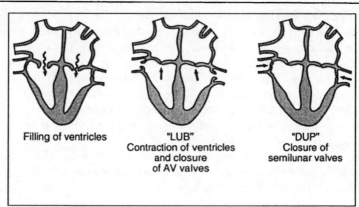

| Filling of ventricles | "LUB"
Contraction of ventricles
and closure
of AV valves | "DUP"
Closure of
semilunar valves |

Fig. 3.1. Heart Sounds.

activity precedes the mechanical, or pumping action, by milliseconds, and is recorded as the electrocardiogram (EKG, ECG).

Calcium-channel blocking agents such as verapamil (Isoptin, Calan), nifedipine (Procardia) and diltiazem (Cardizem), used in the treatment of coronary artery disease and hypertension, block the influx of calcium during repolarization and slow the heart-rate and force of contraction (Fig. 3.2).

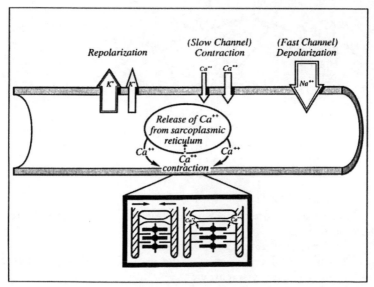

Fig. 3.2. Depolarization of Heart Muscle.

The Conducting System of the Heart

Certain heart muscle fibers depolarize faster than others and constitute the conducting system of the heart. The sino-atrial node (SA node) in the right atrium is the first area to depolarize and sets the heart-rate at about 70 beats per minute. This is the pacemaker. Depolarization spreads throughout the atria (atrial depolarization) and creates the first wave of the EKG, the P-wave. The atrioventricular node (AV node), lying at the interatrial septum, depolarizes and the wave spreads down the interventricular septum to the ventricles. The ventricles depolarize and create the QRS-wave on the EKG. Contraction of the ventricles then takes place (systole). The T-wave is repolarization of the ventricles (the wave for atrial repolarization is masked by the QRS-complex) (Figs. 3.3, 3.4).

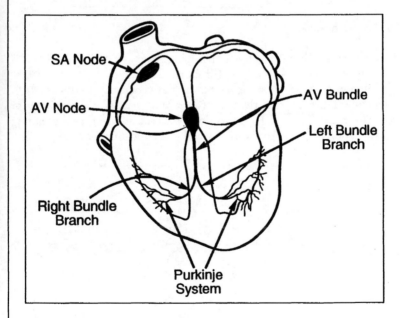

Fig. 3.3. Conducting System of the Heart.

The Heart as a Pump

The ensuing sequence of mechanical events follows the electrical activity of one heart-beat:

1. Diastole: relaxation as the ventricles fill.
2. Atrial systole: contraction of the atria. Blood moves through the AV valves into the ventricles.
3. Systole: contraction of the ventricles. High pressure closes the AV valves (1st heart sound).

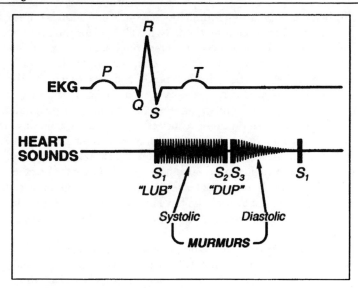

Fig. 3.4. Electrocardiogram and Heart Sounds.

4. Ventricular ejection: blood is forced out the aorta and pulmonary artery.
5. Early diastole: as the pressure lessens after ejection, the pulmonary and aortic valves close (2nd heart sound).

Central Regulation of the Heart

The autonomic nervous system regulates the heart (and other internal organs such as the eye, vessels, lungs, GI tract, bladder and kidney). The system is divided into two branches, the main functions of which are essentially opposites: the sympathetic division ("fight or flight" response) secretes norepinephrine at the synapse (adrenergic); the parasympathetic, or "vegetative" division, maintains normal body functions and secretes acetylcholine at the synapse (cholinergic). The central nervous system (brain and spinal cord) controls autonomic responses.

The medulla and part of the pons control the heart rate and blood pressure. The vasomotor center is a part of a primitive inner core, the reticular formation, running through the brainstem and upper spinal cord. Stimulation of one part of the vasomotor center (sympathetic) causes an increase in heart rate and vasoconstriction, raising the blood pressure. Stimulation of another part causes inhibition of vasoconstriction, resulting in vasodilation and a fall in blood pressure. Stimulation of a third part (parasympathetic) causes a decrease in heart rate by way of the vagus nerve (the vasomotor center is discussed in more depth in Chapter 5).

The heart has a built-in rhythmicity. Nervous stimulation is not needed for contraction. If removed from the body, it will depolarize and contract for a long time. The autonomic nervous system modifies the rate and force of contraction. Receptors for sympathetic nerves are divided into two types, alpha and beta. Beta is subdivided into beta-1 and beta-2 (actually, this is a simplification for clinical purposes—many receptor subtypes exist). Beta-1 and beta-2 receptors are found in many organs and one type predominates. Alpha receptor stimulation causes contraction of vascular smooth muscle and vasoconstriction. Beta-1 receptors predominate in the heart. Beta-2 receptors predominate in bronchial smooth muscle. Stimulation of beta-2 receptors causes bronchodilation. Sympathetic stimulation of the heart releases norepinephrine which stimulates beta-1 receptors and causes an increase in heart rate (chronotropic action) and force of contraction (inotropic action); stimulation of parasympathetic fibers from the vagus nerve decreases the heart rate.

Beta-blocking agents such as propranolol (Inderal), nadolol (Corgard), metaprolol (Lopressor) and atenolol (Tenormin) slow the heart rate and force of contraction by blocking beta-1 receptors. Some beta-2 blocking also occurs (bronchoconstriction) with older drugs. Most of the newer agents are more beta-1 selective.

Inspection and Palpation

In many individuals, the heart-beat may be observed or palpated at about the 5th intercostal space, 7-9 cm left of the midsternum. This is the apical impulse (location of the apex of the heart). With the palm of the hand, one may feel the **apical impulse**, which is usually the point of maximum impulse, or PMI. The strength of the PMI is increased and the location varies in various conditions, such as exercise, emotions, left and right ventricular hypertrophy and hyperthyroidism. Sometimes a vibration (or thrill) is palpated on the chest wall (Fig. 3.5).

Auscultation of the Heart

The stethoscope is placed on the left chest and one listens to the rate, rhythm (regular, irregular) and for abnormal heart sounds (extra sounds, murmurs). In the obese male, heart sounds are best heard with the diaphragm on the left pectoralis major muscle or on the sternal borders at the left or right 2nd interspaces. In the adult female, auscultation is performed by lifting the breast and placing the stethoscope on the chest wall, or placing the diaphragm on the pectoralis major, as in the obese male. The heartbeat is best heard in the quietly supine patient. The heart rate is referred to as the apical rate. Do **not** auscultate the heart over clothing (Fig. 3.6).

Rates are classified as normal (normal sinus rhythm, NSR), fast (tachycardia—100 or more) or slow (bradycardia—60 or less). A normal sinus

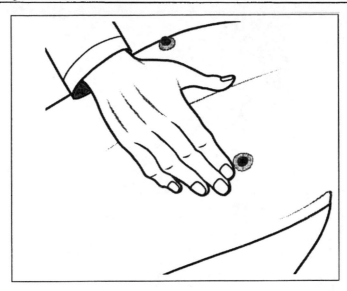

Fig. 3.5. Apical Impulse.

rhythm means that the beat originates in the SA node, the pacemaker. Not all bradycardias or tachycardias are abnormal. Some athletes may have normal rates of 50 or less. Anxiety may increase the rate to 120. Rhythms are classified as regular or irregular (arrhythmias, dysrhythmias). Irregular rhythms are further categorized as regularly irregular (constant regular abnormal beats), such as seen with some premature atrial and ventricular contractions, or irregularly irregular (random irregularity), such as seen with atrial fibrillation and premature ventricular contractions. Combinations of abnormal rates and rhythms frequently exist, such as a bradycardia and an arrhythmia (bradyarrhythmia) or a tachycardia and an arrhythmia (tachyarrhythmia).

Common Fast Rates (Tachycardias)

Sinus tachycardia is a common fast rate of 100-180, originating in the SA node. It is seen during stress, anxiety, pain, and when increased circulatory demands require increased cardiac output, as in shock, congestive heart failure and fever. It is also seen in disease processes such as thyroid storm, adrenal crisis, DKA and renal failure. The EKG shows P, QRS and T-waves. **Treatment:** treating the underlying condition (Fig. 3.7).

Paroxysmal supraventricular tachycardia (PSVT) is a sudden increase in heart rate (usually 140-200) caused by an impulse re-entering the AV node. Some fibers of the AV node conduct at different rates. A signal is conducted to the ventricles by some fibers, then a re-entrant signal travels backward through previously unexcited nodal fibers and initiates a new

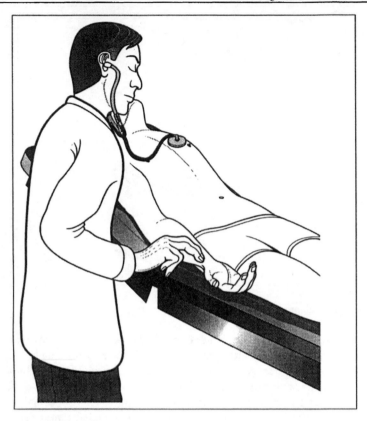

Fig. 3.6. Auscultation of the Heart/Palpation of the Pulse.

impulse, resulting in a sustained tachycardia. PSVT often occurs in healthy individuals. However, the small stroke volume and cardiac output may cause light-headedness. The P-wave is buried in the QRS complex. **Treatment:** in the stable patient, carotid massage is performed while an IV is started—the carotid artery is compressed and massaged with the fingers, stimulating baroreceptors and slowing the heart rate via the vagus nerve. If this is unsuccessful adenosine (Adenocard) is administered as a 6 mg rapid IV bolus, followed by a 20 cc saline flush. If there is no response in 2 minutes, 12 mg is administered. If unsuccessful, diltiazem (Cardizem) 0.25 mg/kg (i.e. 20 mg) is given over 2 minutes, followed 15 minutes later by a second dose if conversion fails (0.35 mg/kg, or 25 mg). In the unstable patient (chest pain, hypotension), after sedation with midazolam 2 mg and morphine sulfate 2 mg, cardioversion at 50J (Figs. 3.8, 3.9).

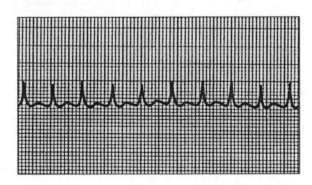

Fig. 3.7. Sinus tachycardia. Reprinted with permission from: Merck, Sharp & Dohm, Division of Merck & Co., Inc.

Fig. 3.8. Paroxysmal Supraventricular Tachycardia (PSVT). Reprinted with permission from: Merck, Sharp & Dohm, Division of Merck & Co., Inc.

Atrial fibrillation with rapid ventricular response resembles paroxysmal supraventricular tachycardia (atrial fibrillation is discussed in the section on arrhythmias). **Treatment:** to control the rate in rapid atrial fibrillation, diltiazem is administered at 20 mg over 2 minutes, followed in 15 minutes by 25 mg in 2 minutes if the first dose is ineffective. In the unstable patient, cardioversion is performed at 100 joules after appropriate sedation (Figs. 3.8, 3.10).

Atrial flutter is a tachycardia of about 150 originating from an ectopic atrial focus depolarizing at 250 to 350 beats per minute. It is usually caused by a reentry mechanism similar to that which causes PSVT. In contrast to PSVT, it is often associated with heart disease. Symptoms may include chest pain, palpitations and light-headedness. The EKG shows "sawtooth" flutter waves preceding each QRS-complex (often 2 flutter waves precede each QRS-complex—2:1 AV block). **Treatment:** after appropriate sedation (see PSVT), cardioversion at 50 J (Figs. 3.10. 3.11).

Ventricular tachycardia (V-tach) is a life-threatening rapid rate (150-200) originating from an ectopic focus or foci in the ventricles. Common causes are ischemic heart disease and myocardial infarction. Symptoms may include

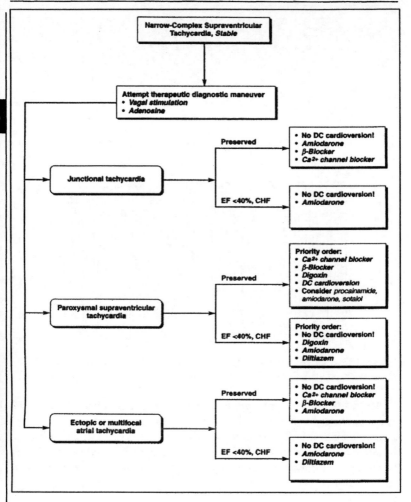

Fig. 3.9. Supraventricular Tachycardia Algorithm. Reprined with permission from: Guidelines for 2000 for Cardiopulmonary Resuscitation and Emergency Cardiovascular Care, American Heart Association.

dyspnea and chest pain. The EKG shows wide QRS-complexes. **Treatment:** since V-tach implies impaired cardiac functioning, amiodarone (Cordarone) 150 mg IV is given over 10 minutes, then every 10 minutes as needed. Alternatively, lidocaine 0.75 mg/kg may be given IV push every 10 minutes, followed by a 3 mg/minute drip if converted to a normal sinus rhythm (pulseless V-tach is treated as V-fib—see Chapter 8) (Figs. 3.12, 3.13).

Control of Rate and Rhythm

Atrial fibrillation/ atrial flutter with • *Normal heart* • *Impaired heart*	1. Control Rate		2. Convert Rhythm	
	Heart Function Preserved	**Impaired Heart EF <40% or CHF**	**Duration <48 Hours**	**Duration >48 Hours or Unknown**
Normal cardiac function	**Note:** *If AF >48 hours' duration, use agents to convert rhythm with extreme caution in patients not receiving adequate anticoagulation because of possible embolic complications.* *Use only 1 of the following agents (see note below):* • Calcium channel blockers (Class I) • β-Blockers (Class I) • For additional drugs that are Class IIb recommendations, see Guidelines or ACLS text	*(Does not apply)*	**Consider** • DC cardioversion *Use only 1 of the following agents (see note below):* • Amiodarone (Class IIa) • Ibutilide (Class IIa) • Flecainide (Class IIa) • Propafenone (Class IIa) • Procainamide (Class IIa) • For additional drugs that are Class IIb recommendations, see Guidelines or ACLS text	• **NO DC cardioversion!** • **Note:** *Conversion of AF to NSR with drugs or shock may cause embolization of atrial thrombi unless patient has adequate anticoagulation.* • Use antiarrhythmic agents with extreme caution if AF >48 hours' duration *(see note above).* *or* ***Delayed cardioversion*** Anticoagulation × 3 weeks at proper levels • Cardioversion, *then* • Anticoagulation × 4 weeks more *or* ***Early cardioversion*** • Begin IV heparin at once • TEE to exclude atrial clot *then* • Cardioversion within 24 hours *then* • Anticoagulation × 4 more weeks
Impaired heart (EF <40% or CHF)	*(Does not apply)*	**Note:** *If AF >48 hours' duration, use agents to convert rhythm with extreme caution in patients not receiving adequate anticoagulation because of possible embolic complications.* *Use only 1 of the following agents (see note below):* • Digoxin (Class IIb) • Diltiazem (Class IIb) • Amiodarone (Class IIb)	**Consider** • DC cardioversion *or* • Amiodarone (Class IIb)	• **Anticoagulation** as described above, followed by • **DC cardioversion**

3

Fig. 3.10. Atrial Fibrillation and Flutter Algorithm. Reprined with permission from: Guidelines for 2000 for Cardiopulmonary Resuscitation and Emergency Cardiovascular Care, American Heart Association.

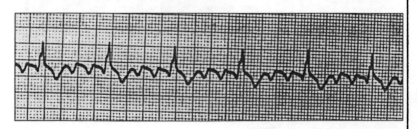

Fig. 3.11. Atrial Flutter. Reprinted with permission from: Conway, A Pocket Atlas of Arrhythmias. © 1974 Year Book Medical Pub.

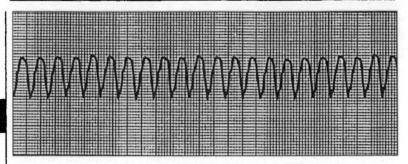

Fig. 3.12. Ventricular Tachycardia. Reprinted with permission from: Merck, Sharp & Dohm, Division of Merck & Co., Inc.

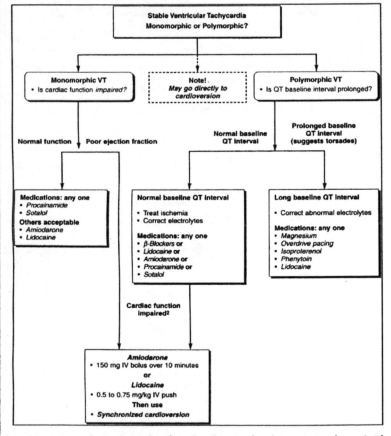

Fig. 3.13. Ventricular Tachycardia Algorithm. Reprined with permission from: Guidelines for 2000 for Cardiopulmonary Resuscitation and Emergency Cardiovascular Care, American Heart Association.

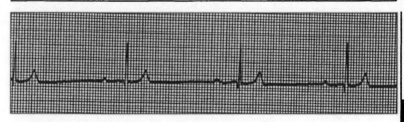

Fig. 3.14. Sinus Bradycardia. Reprinted with permission from: Advanced Cardiac Life Support, ©1997-99 American Heart Association.

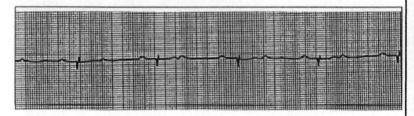

Fig. 3.15. Second Degree Heart Block. Reprinted with permission from: Merck, Sharp & Dohm, Division of Merck & Co., Inc.

Common Slow Rates (Bradycardias)

A common slow rate is sinus **bradycardia**, with a regular rhythm below 60 beats, originating in the SA node. It is seen in physically fit individuals, but also in those on digoxin, beta-blockers, calcium-channel blockers, and in patients with cardiac disease, including myocardial infarction. **Treatment:** for symptomatic patients (obdundation, hypotension), atropine is administered at 1 mg IV every 5 minutes to a maximum of .04 mg/kg. If atropine is unsuccessful, external (transcutaneous) pacing is performed until a transvenous pacemaker can be placed. In cases of **2nd and 3rd degree AV blocks**, where some or all of the fibers of the conducting system are blocked because of disease, initial therapy consists of placing an external pacemaker, followed by a transvenous and then permanent pacemaker (Figs. 3.14, 3.15, 3.16).

Common Arrhythmias

A **sinus arrhythmia** originates in the SA node. The heart rate increases during inspiration and slows during expiration. It occurs normally in children and adolescents and disappears later in life. **Treatment:** none (Fig. 3.17).

Atrial fibrillation is a common malady in the elderly, associated with coronary artery disease, hypertension, hyperthyroidism and rheumatic heart disease. The irregular rate ranges from about 70-100. Multiple areas of the atria depolarize and contract, resulting in muscle quivering. Instead of a

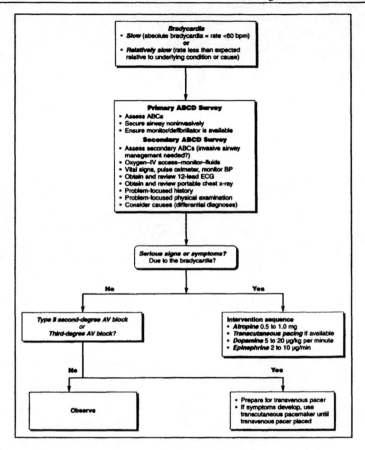

Fig. 3.16. Bradycardia Algorithm. Reprinted with permission from: Guidelines for 2000 for Cardiopulmonary Resuscitation and Emergency Cardiovascular Care, American Heart Association.

normal atrial depolarization (the P-wave), the EKG shows fibrillatory waves accompanied by irregular QRS-complexes. **Treatment:** if the condition is recent (<48 hours), amiodarone (Cordarone) 150 mg IV over 10 min, or ibutilide (Corvert) 0.1 mg/kg IV over 10 min), is effective. If the condition is old and the rate normal, because of the high risk for embolization anticoagulation is begun with coumadin (target INR of 2.5) or aspirin for 3 weeks, depending on risk factors and age. This is followed by electrical or pharmacologic cardioversion (Fig. 3.18).

Premature supraventricular contractions are extra beats originating from either the atria (premature atrial contractions—PAC's) or the AV node (junctional premature beats). They occur in patients with and without heart disease. The EKG shows a premature P-wave in the first, and no P-wave in the latter. **Treatment:** usually none is required (Fig. 3.19).

Premature ventricular contractions (PVC's) are extra beats originating from a single focus (unifocal) or different foci (multifocal) in the ventricle. Many older citizens have occasional PVC's. The condition is aggravated by caffeine, smoking, stress and heart disease. Some cardiac drugs may cause PVC's. The EKG shows a wide QRS without a P-wave, and the complexes are different in configuration from the normal QRS. **Treatment:** efforts should be undertaken to alleviate the underlying cause(s) (Fig. 3.20).

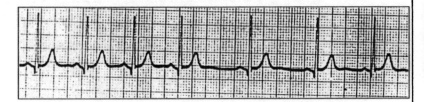

Fig. 3.17. Sinus Arrhythmia. Reprinted with permission from: Conway, A Pocket Atlas of Arrhythmias. © 1974 Year Book Medical Pub.

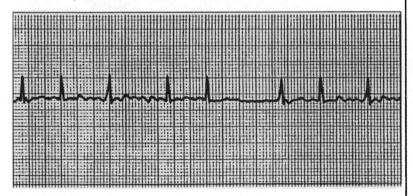

Fig. 3.18. Atrial Fibrillation. Reprinted with permission from: Merck, Sharp & Dohm, Division of Merck & Co., Inc.

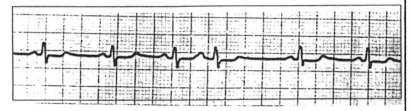

Fig. 3.19. Premature Supraventricular Contraction. Reprinted with permission from: Conway, A Pocket Atlas of Arrhythmias. © 1974 Year Book Medical Pub.

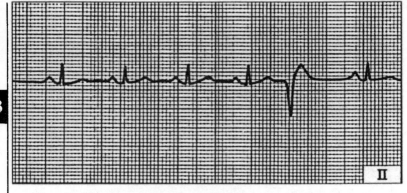

Fig. 3.20. Premature Ventricular Contraction (PVC). Reprinted with permission from: Merck, Sharp & Dohm, Division of Merck & Co., Inc.

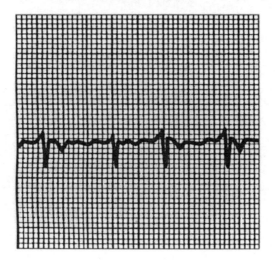

Fig. 3.21. Wandering Pacemaker. Reprinted with permission from: Merck, Sharp & Dohm, Division of Merck & Co., Inc.

Multifocal atrial tachycardia (MAT), or "wandering pacemaker", is seen in elderly patients with chronic obstructive lung disease. In addition to the SA node, two or more different areas of the atrium act as pacemakers (ectopic foci). The EKG shows P-waves of varying morphology and changing PR intervals. **Treatment:** oxygen and bronchodilators. In those with fast rates, magnesium sulfate 2 g IV over 1 minute is sometimes effective (Fig. 3.21).

Special Cases

Blood/Fluid Loss

Abnormal **orthostatic vital signs** are initial indicators of significant blood/body-fluid loss. After about a liter deficit (about 20% of body fluid) an increase in heart rate occurs to compensate for the decreased volume. A patient has orthostatic tachycardia if the heart-rate increases by 30 or more beats per minute or the person becomes light-headed from supine to standing. Only when about 30% of blood-body-fluid loss occurs—about 2 liters—does the systolic pressure begin to drop. An exception to tachycardia with blood loss is sometimes seen. Bradycardia may occur (**paradoxical bradycardia**) because of stimulation of afferent vagal fibers in the left ventricle from cardiac contraction around a reduced blood volume (orthostatic vital signs and paradoxical bradycardia are discussed in detail in Chapter 5).

Doppler Pulse

If one cannot hear the heart or palpate the pulse, a Doppler or ultrasound device may be used. The Doppler transducer, or flowmeter, is a transmitter and receiver and detects the movement of red blood cells, converting the frequency shift of the reflected ultrasound to an audible signal. Acoustic gel is applied. After contact with the skin, the probe is angled in different directions over the artery until an optimum sound is heard with earphones or speaker (Fig. 3.22).

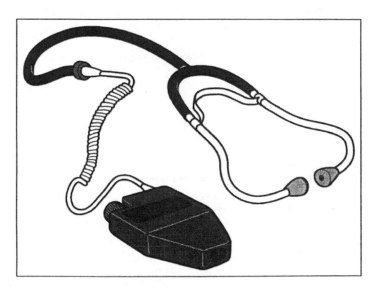

Fig. 3.22. Doppler Stethoscope.

Fetal heart tones (FHTs) may be heard with a regular stethoscope after about 18-20 weeks and with a Doppler stethoscope after about 12 weeks. The mother's heart rate is auscultated prior to the Doppler to avoid confusion. After application of a conducting gel, it is pressed firmly against the abdominal wall. Normal FHTs range from 120-160 beats per minute. Above 160 or below 120 requires urgent obstetric consultation. If fetal bradycardia is present (indicating fetal distress) the mother is placed in the left lateral decubitus position and supplemental oxygen is administered.

Abnormal Heart Sounds

Several normal and abnormal heart sounds are common and may be recognized:

Splitting of the 1st sound. One may hear both AV valves close separately. Usually splitting of the first sound has little clinical significance.

Splitting of the 2nd sound. The aortic valve closes before the pulmonary, sometimes heard in inspiration. If it is heard during expiration it may indicate heart disease.

Third sound. This is a weak sound, heard occasionally after the 2nd and caused by distention of the ventricles during filling. It is loud in heart failure because of overfilling of the failing ventricle. Three sounds are heard in cardiac failure and resemble a galloping horse (gallop rhythm).

Fourth sound. This sound precedes the first and is caused by vibrations and decreased compliance of the left ventricle. It is sometimes heard in myocardial infarction.

A **click** may be heard after the 1st or before the 2nd sound, indicating a normal opening of the semilunar valves, aortic valve disease, pulmonic disease or mitral valve prolapse.

An **opening snap** is sometimes heard after the 2nd sound. It is caused by the opening of a narrowed mitral valve.

A **friction rub**, often heard in infectious pericarditis, is a squeaky or scratchy sound caused by the rubbing together of the dry epicardium against the parietal pericardium.

A **heart murmur** is the "swooshing" sound of blood heard before, during or after (or all three) the heart sounds. It may originate from several mechanisms: 1. increased velocity of blood flow, as in exercise, 2. normal velocity with lessened viscosity, as in anemia, 3. obstruction to flow, as in valvular disease, 4. flow into a dilated chamber, as in an aortic aneurysm and 5. flow through an abnormal opening, as in a congenital heart defect. It is commonly heard in valves that are damaged and do not open properly (narrowing, or stenosis), or close properly, letting blood back up through the valve (regurgitation, or insufficiency).

Murmurs occurring after the first heart sound are systolic. Those occurring after the second heart sound are diastolic. The loudness of the murmur from grade 1 (barely heard) to grade 6 (loudest) and its location is recorded (see Fig. 3.4).

Examples
1. In the setting of an acute myocardial infarction, a new systolic murmur may signal papillary muscle dysfunction or ventricular septal rupture: 3/6 systolic murmur at apex.
2. One may hear the diastolic murmur of aortic regurgitation in a dissecting aortic aneurysm: 5/6 diastolic murmur right sternal border (RSB).

The Pulse

Evaluation of the Pulse

Blood forced into the aorta during systole sets up a pressure wave that travels down the arteries. The wave expands arterial walls. The expansion wave is palpated with the fingertips as the pulse. In contrast to the heart rate, where two sounds are heard with each beat, one beat is felt with the pulse. Palpation is done with the tips of the first two fingers, not the fatty parts, since the digital arteries for each finger anastomose at the fingerpad and using the fatty parts may result in the examiner's own pulse being recorded.

The heart rate may differ from the pulse rate. This is a **pulse deficit**, seen in atrial fibrillation and occasionally in premature ventricular contractions. It occurs in fast rates when some ventricular contractions fail to generate a palpable pulse. One beat is so close to another that the ventricle does not have time to fill and not enough blood is available to produce a pulse wave in the artery. The pulse rate is thus lower than the heart rate. It is discovered when the heart rate is auscultated and the radial pulse is palpated, or when the pulse rate differs from the rate on the cardiac monitor. It is seen in arteries removed from the heart, such as the radial.

Trauma patients and those suspected of having critical conditions such as myocardial infarction, dissecting aortic aneurysm and acute abdominal aneurysm should have pulses assessed in all extremities.

Although various pulse magnitudes and contours exist (i.e., pulsus bigeminus, pulsus bisferiens, Corrigan or water-hammer pulse, etc.), demonstrated by the sphygmograph, the usefulness of these as vital sign parameters is weak. The possible exceptions are pulsus alternans and pulsus paradoxus.

Pulsus alternans is an alternating weak and strong pulse. It is seen in advanced heart failure.

A **paradoxical pulse (pulsus paradoxus)** is an exaggeration of the normal decrease in amplitude of the pulse during inspiration. During inspiration, vessels of the lungs increase in size because of increased negative pressure in the thorax. Blood collects in the lungs, and the stroke volume decreases.

Expiration has the opposite effect. Kussmaul described the condition in 1873 after treating several patients with pericardial effusion. The pulse decreased during inspiration (and in some cases disappeared). However, the heart was obviously still beating, so Kussmaul named the condition "der paradoxe Puls" (see Chapter 1).

A paradoxial pulse is seen when cardiac output is blocked, as in **cardiac tamponade**, but also when lung compliance is decreased, as in COPD. A blood-pressure apparatus was not yet invented in 1873 (Korotkoff first used the Riva-Rocci cuff in 1905) and a link between pulse and blood pressure was not made until Gauchat and Katz at Western Reserve University did so in 1924. Thus, although pulsus paradoxus currently is considered a blood pressure sign, it is actually a pulse sign (the blood-pressure counterpart of the sign is described in Chapter 5). In a busy emergency setting where nuances of change in blood-pressure readings are difficult to detect, reversion to Kussmaul's palpation of an artery is more useful. As an example, in the trauma setting when a penetrating injury to the chest is present, gradual disappearance of the radial pulse on inspiration may herald an impending cardiac tamponade.

Peripheral Pulses

The artery commonly used for pulse-taking is the **radial**, lying lateral to the flexor carpi radialis tendon on the distal radius (Fig. 3.23). It is sometimes difficult to find.

The second most useful is the **brachial**, because of blood pressure taking. Its location sometimes surprises people (Fig. 3.24). It is more easily palpable medial, not lateral, to the biceps tendon and superior to, not in, the antecubital fossa (cubital is forearm; antecubital is volar forearm. The differences have become obscured and the two terms are often used synonymously). In the antecubital fossa, the brachial artery divides into the radial and ulnar arteries. The ulnar goes deep and the radial crosses the biceps tendon and runs laterally down the forearm. If the stethoscope is placed in the antecubital fossa, the blood pressure is being measured in the proximal portion of the radial artery, not the brachial. Accurate palpation of the brachial artery alleviates multiple attempts at blood-pressure taking.

The **common carotid** (Fig. 3.25) artery lies deep and slightly anterior to the sternocleidomastoid muscle. One must be careful to lightly palpate the artery, since sustained pressure will activate the baroreceptor mechanism and slow the heart rate. Do not palpate both carotid arteries at the same time or fainting may occur.

The **femoral** artery (Fig. 3.26), the largest of the pulse-taking arteries, is located at the midpoint of the inguinal ligament between the anterior superior iliac spine and the pubic symphysis. It is the more useful for palpation in infants, the obese, the elderly and during cardiopulmonary resuscitation.

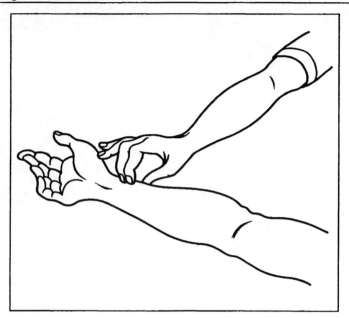

Fig. 3.23. Radial Pulse.

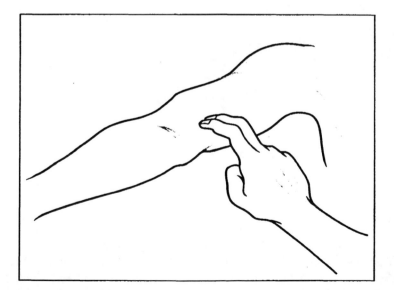

Fig. 3.24. Brachial Artery.

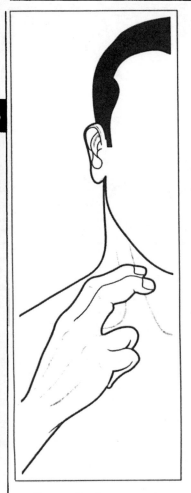

Fig. 3.25. Carotid Artery.

The **popliteal** artery is the continuation of the femoral at the popliteal fossa. It lies deep and medial to the popliteal vein and tibial nerve and is frequently difficult, if not impossible, to find. Searching for it is unnecessary if good femoral and pedal pulses are present. Popliteal palpation evaluates patency when foot arteries are unavailable.

In the foot, the **posterior tibial artery** is the continuation of the popliteal and is sometimes difficult to locate. It lies behind and below the medial malleolus. Often an easier one to find is the **dorsal pedis** on the dorsum of the foot at the junction of the first two extensor tendons (extensor hallucis longus and brevis—hallus: Latin—great toe). It is helpful to mark the area with an "X" for a difficult-to-find dorsal pedis pulse (or any other) (Figs. 3.27, 3.28).

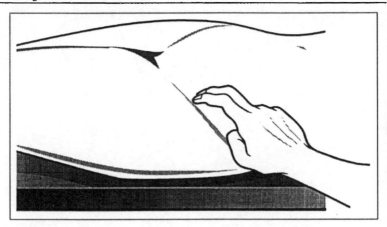

Fig. 3.26. Femoral Artery.

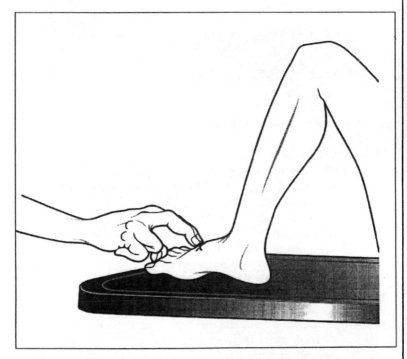

Fig. 3.27. Dorsal pedis artery.

Fig. 3.28. Posterior tibial artery (right leg).

Practical Points
- Record the rate and rhythm (regular, irregular), as well as the quality and strength of the pulse (weak, strong, thready). Examples:
  1. L radial—54, reg, weak.
  2. R femoral—130, irreg, thready.
- Always auscultate the heart and palpate the pulse. The rates may differ (pulse deficit). Example: HR—120, L radial P—66.
- Never be satisfied with one set of vitals.
- In a patient with a possible vascular event, such as a dissecting thoracic aneurysm, take pulses in all extremities.
- Sometimes in the obese patient and others a radial and brachial pulse cannot be felt, and one cannot hear a heart beat. If the patient is comatose, palpate the femoral. It is the easiest to find (in the elderly the carotid is sometimes stenosed and difficult to find). Use a Doppler if necessary.

- Never rely on a monitor or any electronic device for the heart rate. Murphy's Law will exert its inexorable effect and although a normal sinus rhythm will be showing on the monitor the patient will have no pulse and will be moribund or dead.
- Do not auscultate the heart over clothing.

References

1. American Heart Association and the International Liaison Committee on Resuscitation (ILCOR): Guidelines 2000 for cardiopulmonary resuscitation and emergency cardiovascular care. Baltimore: Lippincott, Williams & Wilkins, 2000.
2. Barach P. Pulsus paradoxus. Hosp Phys 2000; 36:49.
3. Bolton E. Disturbances of cardiac rhythm and conduction. In: Tintinalli et al, eds. Emergency Medicine: A Comprehensive Study Guide. New York: McGraw-Hill, 2000.
4. Conway N. A Pocket Atlas of Arrhythmias. Chicago: Year Book Medical Pub, 1974.
5. DeGowan R et al. Bedside diagnostic examination, New York: Macmillan Pub. Co., 2000.
6. Hoffman B. Adrenoceptor-activating drugs. In: Katzung B, ed. Basic and Clinical Pharmacology. Norwalk: Appleton & Lange, 1989.
7. Koziol-McLain J et al. Orthostatic vital signs in emergency department patients. Ann Emerg Med 1991; 20:6.
8. Kussmaul A. Ueber schweilige Mediastino-Pericarditis und den paradoxen Puls. Berl Klin Wochenschr 1873; 10:37-39.
9. Lewinter J et al. Vital sign measurement procedures. In: Roberts J, Hedges J, eds. Clinical Procedures in Emergency Medicine. Philadelphia: WB Saunders, 1998.
10. McGregor M. Pulsus paradoxus. N Engl J Med 1979; 301:480.
11. Miles W et al. Arrhythmias. In: Andreoli et al, eds. Cecil Essentials of Medicine, Philadelphia, 1997.
12. Niemann J. The cardiomyopathies, myocarditis and pericardial disease. In: Tintinalli et al, editors: Emergency Medicine: A Comprehensive Study Guide. New York: McGraw-Hill, 2000.
13. O'Rourke M et al. The Arterial Pulse. Philadelphia: Lea & Febiger, 1992.
14. O'Rourke R. The measurement of systemic blood pressure; normal and abnormal pulsations of the arteries and veins. In: Hurst J et al. The Heart. New York: McGraw-Hill, 1990.
15. Shabetai R et al. Pulsus paradoxus. J Clin Invest 1965; 44:11.
16. Stewart J. Clinical Anatomy and Physiology for the Angry Health Professional. Miami: MedMaster Inc., 2001.
17. Volgman A. Managing atrial fibrillation: What is the value of adding aspirin to warfarin therapy? J Crit Ill 2000; 15:185.
18. Wo C et al. Unreliability of blood pressure and heart rate to evaluate cardiac output in emergency resuscitation and critical illness. Crit Care Med 1993; 21:2.
19. Yealy D, Delbridge T. Dysrhythmias. In: Rosen P et al, eds. Emergency Medicine: Concepts and Clinical Practice. St. Louis: Mosby Year Book, 1998.
20. Zide R, Tsapatsaris N. Use of anticoagulation: Addressing atrial fibrillation and deep venous thrombosis. Res & Staff Phys 1999; 45:21.

CHAPTER 4

Vital Sign #3: Respiration

The name of the vital sign proposed by Edward Seguin in 1866 was "respirations". Over the years the name changed to "respiratory rate" (RR), the original emphasis of the sign. Today in many charts it is back to "respiration(s)", indicating a more thorough evaluation.

Respiration is the more critical of the vital signs, since the heart and brain require a definite amount of oxygen in order to function. In emergencies, the airway is first addressed, then breathing, then circulation—the ABCs of resuscitation. Apnea for more than 7-10 minutes usually means irreversible brain damage. Exceptions exist, including infants, and isolated cases of hypothermia and drowning.

A protocol was developed a few years ago for the Advanced Trauma Life Support (ATLS) course also, relevant to the nontrauma patient: **Primary Survey** (ABCDs), **Resuscitative Measures** and **Secondary Survey** (history and physical exam).

The **Airway** is opened with jaw thrust or chin lift, oxygen is administered, pulse oximetry is monitored, the airway is secured by intubation or cricothyrotomy and the cervical spine is cleared. **Breathing** is assisted with bag-valve-mask or ventilator, tension pneumothorax is decompressed by needle, a chest tube is placed for pneumothorax or hemothorax and pulmonary edema is treated. **Circulation**: IV access is obtained, a cardiac monitor is placed, dysrhythmias and shock are treated, and **Disability**: a brief neurological exam is assessed. If decreased level of consciousness is present, a chemstrip glucose is obtained (or if unavailable glucose is given) and thiamine and naloxone are administered. The above protocol appears as Figure 8.15.

A thorough discussion of **Disability** is found in Chapter 6, and **Resuscitation** in Chapter 8.

Anatomy and Physiology

The Mechanics of Breathing

During inspiration, the external intercostal muscles contract, lifting the lower ribs up and out. The diaphragm moves down, increasing the volume of the pleural cavity. Elastic fibers in the alveolar walls stretch, and the air sacs of the lungs expand. During expiration, the external intercostals and diaphragm relax. In diseases such as asthma and COPD **accessory muscles**

Vital Signs and Resuscitation, by Joseph V. Stewart. ©2003 Landes Bioscience.

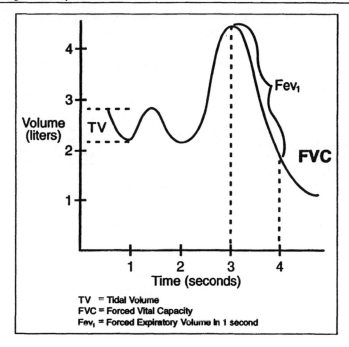

TV = Tidal Volume
FVC = Forced Vital Capacity
Fev₁ = Forced Expiratory Volume in 1 second

Fig. 4.1. Lung Volumes.

of respiration may be used: inspiration is assisted by the sternocleidomastoids and scalene muscles; expiration is aided by the internal intercostals and abdominal muscles.

Lung Volumes and Pulmonary Function Testing

Breathing in and out creates volumes that can be measured. In earlier times this was done with a spirometer, a revolving drum with a writing lever that recorded lung volumes on graph paper. Today a computer program is used. The patient blows into a small hand-held device (flow sensor) and the results are displayed on a computer screen. Key elements are the forced vital capacity (FVC)—the maximum volume one can forcibly expel after a maximum inspiration (i.e., 4800 ml) and the forced expiratory volume in 1 second (FEV-1)—the amount of air forcibly expelled in 1 second (i.e., 83%). The tidal volume (TV) is the amount of air moved during normal respiration. An asthmatic, for example, often has a normal TV and FVC but decreased FEV. In emphysema all three are usually decreased. The peak expiratory flow rate (PEFR), or **peak flow**, is a method of respiratory evaluation that can be done quickly in the emergency setting. A small hand-held device (peak flow meter) is used. The maneuver is similar to the FVC but recorded in liters per minute (i.e., 550 L/min). A reading below 200 L/min usually indicates respiratory compromise (Figs. 4.1, 4.2).

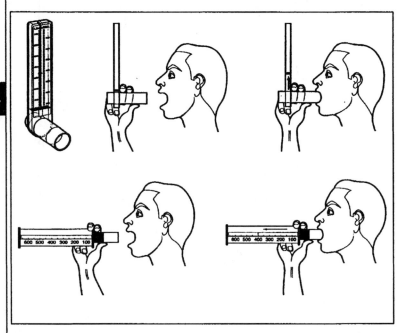

Fig. 4.2. Peak Flow Meters.

The Normal Respiratory Rate

The respiratory rate in adults is 12-18 breaths per minute. In the newborn it is about 40 and decreases to adult values at age 18. When a person realizes that the respiratory rate will be observed, he becomes self-conscious and begins to breathe in an odd fashion. Thus, the usual practice is to examine the heart or pulse and observe respirations without mentioning it.

The Physiology of Respiration

Oxygen from air enters the lungs and diffuses through the alveolar and capillary membranes into the bloodstream. The pulmonary veins return oxygen-rich blood to the left side of the heart, where it is pumped to the rest of the body via the aorta. Oxygen is transported in the blood as oxyhemoglobin in the red cells.

In the tissues, red cells move into the capillaries. At the arteriole end of the capillary, oxygen diffuses through the red cells, then through the capillary membrane into the tissue fluid. It then diffuses through the tissue cell membrane to be used as fuel for cellular metabolism.

Carbon dioxide moves out of tissue cells in the reverse direction into red cells, where most is converted to bicarbonate ion (HCO_3^-). Bicarbonate is transported in the plasma. The large veins bring oxygen-poor blood to the right side of the heart where it is transported via the pulmonary arteries to the lungs. In the lungs the process is reversed in the alveoli, and carbon dioxide is blown off (Fig. 4.3).

$$H^+ + HCO_3^- \longleftrightarrow (H_2CO_3) \longleftrightarrow H_2O + CO_2 \text{ (BLOWN OFF)}$$

Fig. 4.3. Bicarbonate buffer equation.

Two tools are available for the analysis of the pH, oxygen and carbon dioxide content of blood. The **pulse oximeter** is a computer and probe consisting of 2 photodiodes and photodetector that attaches to the fingertip and measures the **oxygen saturation** of arterial blood. **Blood gas analyzers**, using blood gas and pH electrodes, measure the partial pressures of oxygen, carbon dioxide and pH of blood. Other values, such as oxygen saturation (SaO_2) and bicarb level (HCO_3^-) are calculated. Normal arterial blood gas values are as follows (each lab may differ slightly):

1. pO_2 = 85—105 mmHg
2. pCO_2 = 35—45 mmHg
3. pH = 7.35—7.45
4. HCO_3^- = 21—26 meq/l
5. SaO_2 = 95—100%

The normal oxygen saturation is between 97 and 100%. Below 94% represents hypoxia. Severe hypoxia is present with an oxygen saturation of 90%. As seen in the **oxyhemoglobin dissociation curve**, an oxygen saturation of 90% represents a p02 of only 60 mm Hg (Fig. 4.4).

Regulation of Respiration

Respiration is controlled by the r**espiratory center**, nerve cells in the reticular formation of the pons and medulla. Impulses from the cerebral cortex modify respirations, as do changes in the oxygen content, carbon dioxide content and the pH of blood. Cells sensitive to these changes are **chemoreceptors**, located in the medulla, the arch of the aorta (aortic bodies) and junction of the internal and external carotid arteries (carotid bodies). Low oxygen, high carbon dioxide or low pH activates the chemoreceptors and causes the respiratory rate to increase. Low carbon dioxide or a high pH has the opposite effect. Impulses from the aortic and carotid bodies travel to the respiratory center in the brainstem via the vagus and glossopharyngeal nerves (Fig. 4.6).

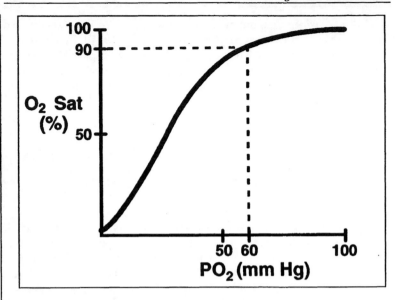

Fig. 4.4. Oxyhemoglobin-Dissociation Curve.

Acid/Base Conditions

The bicarbonate buffer equation illustrates the effects of various physiological conditions. Carbonic acid is transiently formed and carbon dioxide is blown off during expiration (Fig. 4.5).

$$H^+ + HCO_3^- \longleftrightarrow (H_2CO_3) \longleftrightarrow H_2O + CO_2 \uparrow \text{(BLOWN OFF)}$$

Fig. 4.5. Bicarbonate Buffer Equation.

In **metabolic acidosis**, a common problem, the body produces an increased amount of acid (H^+), which combines with bicarbonate to form CO_2 and water. In compensated metabolic acidosis, the body adjusts to keep the pH within normal limits. In uncompensated metabolic acidosis the body is unable to cope with the acid load and the pH begins to fall. The low pH stimulates the respiratory center and the equation shifts to the right. As H^+ continuously combines with HCO_3^-, the HCO_3^- begins to fall. As CO_2 accumulates, the body breathes deeper and/or faster to eliminate it. This is seen in diabetic ketoacidosis. An accumulation of acidic substances in kidney disease or aspirin overdose (impairment of oxidative phosphylation, a

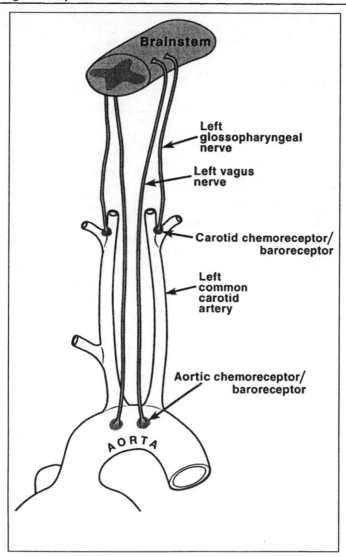

Fig. 4.6. Baroreceptors/chemoreceptors.

major buffer of H⁺) may also cause metabolic acidosis. **Treatment:** The underlying cause is treated. If the pH is less than 7 or the HCO_3^- is becoming depleted, 1 or 2 amps of bicarbonate are given intravenously (Fig. 4.7).

In **metabolic alkalosis**, a rare condition, acid is lost, the equation shifts to the left to restore acid, and ventilation decreases. This is seen occasionally in those who have been vomiting for a long time (depletion of HCl from the

$$H^+ + HCO_3^- \longrightarrow CO_2 \uparrow$$

Fig. 4.7. Metabolic Acidosis.

$$H^+ + HCO_3^- \longleftarrow CO_2$$

Fig. 4.8. Metabolic Alkalosis.

stomach) and in those using diuretics (loss of H^+ from the kidney). **Treatment:** correcting the condition (Fig. 4.8).

Respiratory acidosis is seen in severe asthma, chronic obstructive pulmonary disease (COPD) and in conditions in which ventilation is poor, such as congestive heart failure and pneumonia. The person is unable to blow off CO_2, which accumulates. The equation shifts to the left resulting in rising acidity. The respiratory rate increases in an effort to blow off accumulated CO_2. **Treatment:** increasing ventilation with bronchodilators. Occasionally intubation and assisted ventilation are required (Fig. 4.9).

Respiratory alkalosis occurs as a compensatory reaction to metabolic acidosis and in anxiety reactions (hyperventilation syndrome—see next section). Treatment for hyperventilation is a quiet setting to restore CO_2, and in the case of metabolic acidosis (i.e. ketoacidosis) the underlying condition is treated (Fig. 4.10).

$$H^+ + HCO_3^- \longleftarrow CO_2 \uparrow \text{(BLOCKED)}$$

Fig. 4.9. Respiratory Acidosis.

$$H^+ + HCO_3^- \longrightarrow CO_2 \uparrow$$

Fig. 4.10. Respiratory Alkalosis.

Often two acid-base conditions occur together. In diabetic ketoacidosis, as mentioned, the metabolic acidosis triggers a compensatory respiratory alkalosis. In this case, the HCO_3^- is low, but so is the CO_2. The pH will be close to normal. The person with COPD in respiratory acidosis may also have developed an additional metabolic acidosis. In this case, instead of the HCO_3^- being close to normal, it will fall. In general, if the pH is close to normal, and the CO_2 and/or HCO_3^- are abnormal, one may assume a mixed condition.

Atypical Breathing

Fast Breathing (Tachypnea)

Tachypnea is usually significant at rates above 20. A low oxygen, a high CO_2, or a low pH (or combinations) may cause tachypnea, seen in asthma, pneumonia, CHF, exacerbations of COPD and DKA. Other conditions causing tachypnea are emotional reactions (i.e., hyperventilation), pulmonary embolism, pneumothorax, obesity (increased vessel resistance), pain (increased nervous stimulation), anemia (decreased oxygen) and hyperthyroidism (increased metabolic rate). An increased respiratory rate is also seen with sympathomimetric drugs, as well as aspirin, methanol, ethylene glycol and carbon monoxide poisonings. Treatment is directed at correcting the underlying condition.

Although **hyperventilation** is usually the result of anxiety, life-threatening conditions such as pulmonary embolism, diabetic ketoacidosis and sepsis must first be ruled out. Carbon dioxide is blown off, creating respiratory alkalosis. The person presents with a feeling of shortness of breath, lightheadedness and tingling in the hands, feet and mouth (decreased CO_2 causes cerebral vasoconstriction, reduced cerebral blood flow and paresthesias). The alkalosis causes increased binding of calcium to plasma protein, decreasing the amount of ionized calcium in the bloodstream. The low calcium results in spasms of skeletal muscles (tetany), and the person often arrives in the emergency department in carpal spasm. This reverses as the CO_2 returns to normal. **Treatment:** carbon dioxide is restored by decreasing the respiratory rate in a quiet environment (breathing into a paper bag should be avoided because of the potential for hypoxia).

Slow Breathing (Bradypnea)

A slow respiratory rate is usually significant at a rate of 8 or less per minute. Often this is an emergency and requires immediate therapy. Conditions causing bradypnea are the ingestion of drugs (i.e., alcohol, narcotics, sedative-hypnotics), increased intracranial pressure from trauma and hemorrhage (pressure on the respiratory center), severe respiratory depression (i.e., CO_2 narcosis) and coma from any cause. It is seen in many pre-arrest and end-stage conditions. **Treatment:** assisted ventilation is often required with a bag-valve-mask (BVM). Endotracheal intubation is frequently necessary.

Irregular Breathing

Cheyne-Stokes Breathing
This type of irregular respiratory pattern is observed in terminal situations where tachypnea alternates with apnea. It is seen in severe central nervous system injuries such as stroke, hypertensive encephalopathy, brain swelling from trauma with impending herniation (increased intracranial pressure) and in severe heart failure. The cause is an altered cerebral response to CO_2. Overbreathing is present when the CO_2 is elevated, then apnea occurs to restore the CO_2. In severe heart failure, the sluggish circulation causes a delay and overcorrection of the acid-base status.

Kussmaul Breathing
Described in 1873 by the German physician Adolf Kussmaul, "Lufthunger" or "air hunger" is the deep breathing seen in diabetic ketoacidosis and uremia to blow off carbon dioxide produced in metabolic acidosis. The rate may be slow, regular or fast. Treatment is directed at the underlying cause.

Sleep Apnea
In some obese individuals, drowsy episodes accompanied by snoring and apneic spells occur. This obstructive sleep apnea is caused by one or more anatomic abnormalities. The tongue falls back during sleep and blocks the airway. Treatment involves weight loss, avoidance of alcohol and nasal continuous positive airway pressure (nasal CPAP) at night. Resection of pharyngeal soft tissue may be required.

Abnormal Respiratory Sounds
1. Snoring respirations are sometimes caused by the tongue falling back in the throat, partially obstructing the upper airway. A jaw thrust or chin lift corrects the situation.
2. **Stridor** is the high-pitched sound of air moving through a partially obstructed upper airway.
3. Decreased breath sounds in a portion of a lung (usually the base) may be caused by a pneumothorax, hemothorax or a large pleural effusion.
4. **Rales** (pronounced "rahls", also called crackles) are sounds like tissue paper being squeezed, indicating fluid in the small airways and alveoli.
5. **Rhonchi** are rattling sounds from mucous and fluid in the large airways (bronchi).
6. **Wheezes** are musical sounds produced by air moving through narrowed bronchi and bronchioles.

Labored Breathing (Dyspnea)

Upper Airway

The most common cause of upper airway obstruction is a decreased level of consciousness from any cause. The tongue falls back in the mouth, partially obstructing the airway. Treatment is a jaw thrust or chin lift, and insertion of a nasopharyngeal or oropharyngeal airway (Fig. 4.11).

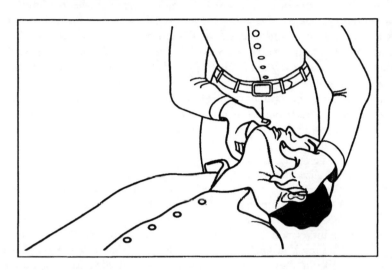

Fig. 4.11. Jaw Thrust.

Signs of upper airway obstruction include snoring respirations, shortness of breath, cyanosis, hoarseness, difficulty in swallowing (dysphagia) or speaking, stridor, coughing, grunting or tachypnea in any combination. In the pediatric population, tachypnea, chest retractions and nasal flaring are often prominent. Causes are foreign bodies, trauma, allergic reactions and infection. These are frequently medical emergencies. Treatment depends on the specific problem. Foreign bodies may be removed manually. With trauma patients, if intubation is not possible, a cricothyrotomy is performed. An allergic reaction involving the upper airway (**angioedema**) or a systemic reaction (**anaphylaxis**) is treated with epinephrine, antihistamines and steroids (see Chapter 5, Anaphylactic Shock).

Lower Airway

Common lower airway problems causing dyspnea are asthma, COPD, pneumonia, pulmonary edema, pulmonary embolism/infarction and pneumothorax (see following section).

Common Examples of Labored Breathing

Asthma

Asthma is an allergic disorder affecting bronchi and bronchioles. Smooth muscle constricts and glands of the bronchi secrete increased amounts of mucous. Air enters the alveoli but leaves with difficulty. The result is wheezing. Wheezing is not critical unless the patient is using accessory muscles to force air out of the lungs. Occasionally in a tiring patient, wheezes diminish and little air is moved, heralding respiratory failure and requiring endotracheal intubation and mechanical ventilation. Tests reflecting the patient's respiratory status are **pulse oximetry** and **peak flow** (see earlier section). **Blood gases** are usually not required (they show a partial respiratory alkalosis—CO_2 is blown off). However, if done, a normal CO_2 in a tiring asthmatic indicates impending respiratory failure. A peak flow of less than 200 L/min after several nebulizer treatments is usually an indication for hospitalization (normal peak flow is over 400 liters per minute).

Treatment:
1. oxygen by cannula or mask,
2. nebulizer therapy: a predominantly beta-2 agent such as albuterol (Ventolin), an anticholinergic bronchodilator such as ipratropium (Atrovent), combinations, or a more beta-2 selective agent such as levalbuterol (Xopinex) is administered,
3. an intravenous steroid such as methylprednisolone (Solu-Medrol) 125 mg is given. In addition to having delayed long-acting anti-inflammatory effects, steroids act synergistically with beta-2 aerosols to abort some of the bronchospasm of asthma.

It is important to remember that, from a vital sign standpoint, a person may have a normal respiratory rate, not be wheezing and be in severe respiratory failure, as signaled by use of accessory muscles, sweating, tiring, deterioration of mental status and movement of little air. In this case endotracheal intubation is required (see respiratory failure, Chapter 8; see also Pediatric Asthma, Chapter 7).

Chronic Obstructive Pulmonary Disease

Chronic obstructive pulmonary disease (COPD) refers to two disease entities sharing characteristics of long term obstruction to air flow: **chronic bronchitis** and **emphysema**. Smoking is often a component in both diseases. In bronchitis, the bronchial mucosa is swollen and red, mucous is secreted by the glands, and the sputum may be green or yellow. In emphysema (Gr: "to puff up") air is trapped in the alveoli because of long-term irritation of the bronchi, and mucous and pus accumulate. When pressure in the alveoli exceeds the elastic limit, they become permanently ballooned-out and nonelastic. This produces the barrel-chested person sometimes requiring use of accessory muscles of respiration to breathe. Wheezes are often heard in both situations. Pulse oximentry usually shows chronic hypoxemia. A peak flow is less useful than in asthma, and is usually measured against the patient's baseline. **Treatment:**

1. low-flow oxygen at 2 L/min (high-flow may abolish the hypoxic ventilatory drive and lead to respiratory arrest), or 28% by Venturi mask,
2. bronchospasm is treated with a beta-agonist such as albuterol 2.5 mg, levalbuterol (Xopinex), or an anticholinergic agent such as iprotropium (Atrovent) 500 μg in 2 ml normal saline by nebulizer,
3. inflammation is treated with a steroid such as methylprednisolone 125 mg IV,
4. an antibiotic (amoxicillin or trimethoprim-sulfamethoxazole) is administered since the exacerbation is usually the result of an infection, and
5. stopping smoking helps dramatically.

Pneumonia

Signs and symptoms of pneumonia are fever, chills, cough, production of rust-colored sputum, chest pain, tachypnea, dyspnea, decreased breath sounds and rales. The CBC shows a leukocytosis and a chest x-ray usually reveals an infiltrate. **Treatment:** viruses require no therapy. A bacterial infection is treated with an appropriate antibiotic based on gram stain or probable etiology.

Pulmonary Edema

Acute pulmonary edema is a life-threatening sequel of congestive heart failure, often triggered by failure to take appropriate medication and sometimes by an acute myocardial infarction. Because of inadequate pumping action of the left ventricle, fluid backs up in the lungs. Cough, orthopnea and chest pain are common symptoms. Anxiety, dyspnea, tachypnea, rales, wheezes, tachycardia with an S-3 gallop rhythm, jugular venous distention (JVD), peripheral edema and diaphoresis may be present. Blood gases show hypoxia and sometimes hypercapnia. A chest x-ray reveals diffuse infiltrates in both lungs. **Treatment:**

1. upright position,
2. high flow oxygen by mask,
3. a diuretic such as furosemide 80 mg IV to remove fluid,
4. nitroglycerine (NTG) 10 μg per minute by intravenous infusion for vasodilation, reducing preload (and some afterload),
5. morphine, 2 mg IV, although controversial, slightly reduces afterload, cardiac work and produces a sedative effect,
6. a systolic pressure <100 mmHg may require the administration of dopamine (5 μg/kg/min).

7. severe hypertension not responding to NTG may require nitroprusside 0.5 µg/kg/min. Intubation is frequently required. First-line therapy may be remembered by the **mnemonic:** L (lasix), M (morphine), N (nitrates), O (oxygen), P (position) (Fig. 4.12).

Pulmonary Embolism/Infarction

In pulmonary embolism, a clot from a pelvic or deep leg vein detaches and travels to the lung. The clot impacts in a branch of the pulmonary artery causing chest pain, dyspnea, tachypnea and sometimes syncope, anxiety, cough and hemoptysis. A large embolus occluding a major branch of the pulmonary artery (infarction) may cause shock and even death. Predisposing factors include a previous pulmonary embolism, deep vein thrombosis (DVT), CHF, MI, obesity, recent surgery, immobilization, trauma, pregnancy and malignancy. Diagnosis is made by lung scan (normal ventilation, with perfusion defects), or a spiral CT scan. If results are equivocal, a pulmonary angiogram is done. Doppler studies of the leg are often positive.

Treatment:

1. oxygen by cannula or mask to maintain an oxygen saturation at 95%,
2. heparin 10,000 unit IV bolus and 1000 units per hour, or low molecular weight heparin such as enoxaparin (Lovenox) at 1 mg/kg subq q 12 hours,
3. IV normal saline,
4. dopamine (5 µg/kg/min) may be required for hypotension. Long term therapy includes an oral anticoagulant such as warfarin (Coumadin).

Note: pulmonary embolism is one of the more missed diagnoses. An increased respiratory rate with some hypoxia is almost always present.

Pneumothorax

A pneumothorax is air between visceral and parietal pleurae. It occurs from rupture of a pulmonary bleb in the lung of the asthmatic/COPDer, in the trauma patient (particularly knife or bullet wound) and sometimes in cancer patients. Symptoms are sharp chest pain and cough. Occasionally dyspnea is present. Unless the pneumothorax is quite small, breath sounds are decreased on one side. A chest x-ray is usually diagnostic. Occasionally, air may compress the mediastinum and vena cavae, resulting in severe respiratory distress, tachycardia and hypotension (**tension pneumothorax**). **Treatment:** close observation for a small nontraumatic pneumothorax. A greater than 20% pneumothorax usually requires a chest tube (tube thoracostomy). For tension pneumothorax, a 14/16-gauge needle/catheter is inserted in the second interspace, midclavicular line, followed by tube thoracostomy (Fig. 4.13; see also Fig. 8.7).

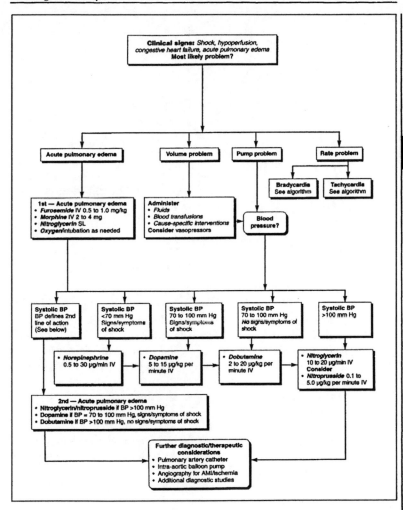

Fig. 4.12. Pulmonary Edema, Hypotension, Shock Algorithm. Reprinted with permission from: Guidelines for 2000 for Cardiopulmonary Resuscitation and Emergency Cardiovascular Care, American Heart Association.

Practical Points

- First, the ABCs of resuscitation are followed (see also Chapter 8).
- The respiratory rate is counted per minute while taking heart rate and blood pressure.
- Note any pattern (i.e., respiratory rate 30, followed by apnea for 10 seconds), the depth of respirations (strong, shallow) and abnormal sounds (labored, snoring, stridor, rales, wheezes).

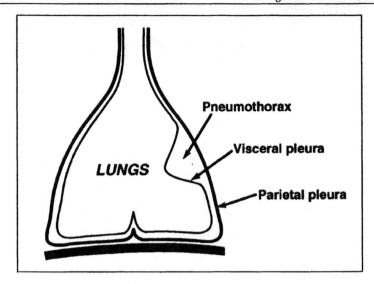

Fig. 4.13. Pneumothorax.

- Indicate if breath sounds are diminished on one side (if so, a needle or chest tube may be immediately required).
- Record use of accessory muscles, and whether the person has to sit up to breathe. Note respiratory distress or dyspnea (mild, moderate, severe).
- Do not auscultate the lungs over clothing.
  - Examples:
  1. RR 12/breath sounds strong and equal bilaterally
  2. RR 6/shallow respirations/moderate resp distress
  3. RR 25/bilateral wheezes/mild distress/use of accessory muscles.
  4. RR 30/bilateral basilar rales
  5. RR 38/shallow resp/accessory muscles/moderate distress
  6. RR 10/labored breathing/no breath sounds on left

References

1. ACC/AHA Task Force. Guidelines for the evaluation and management of heart failure. Circulation 1995; 92:2764.
2. American Heart Association and the International Liaison Committee on Resuscitation (ILCOR). Guidelines 2000 for cardiopulmonary resuscitation and emergency cardiovascular care. Baltimore: Lippincott, Williams & Wilkins, 2000.
3. Baker W et al. Noninvasive assessment and support of oxygenation and ventilation. In: Hedges R. Clinical Procedures in Emergency Medicine. Philadelphia: WB Saunders, 1998.
4. Barrett S. Dyspnea and shortness of breath. In: Rosen P et al. Emergency Medicine: Concepts and Clinical Practice. St. Louis: Mosby Year Book, 1998.

5. Baumann M, Strange C. The clinician's perspective on pneumothorax management. Chest 1997; 112:822.

6. Carpenter L, Verdile V. Arterial blood gas analysis. In: Wolfson A, Paris P. Diagnostic Testing in Emergency Medicine. Philadelphia: WB Saunders, 1996.

7. Crapo R. Pulmonary-function testing. N Eng J Med 1994; 331:1.

8. Cydulka R and Khandelwal S. Chronic obstructive pulmonary disease. In: Tintinalli J et al. Emergency Medicine: A Comprehensive Study Guide. New York: McGraw-Hill, 2000.

9. Expert Panel Report. Guidelines for the diagnosis and management of asthma. Bethesda: National Institutes of Health, 1991.

10. Grifoni S et al. Short term clinical outcome of patients with acute pulmonary embolism. Circulation 2000; 101:2817.

11. LeConte P et al. Prognostic factors in acute cardiogenic pulmonary edema. Am J Emerg Med 1999; 17:329.

12. Lin R et al. Rapid improvement of peak flow in asthmatic patients treated with parenteral methylprednisolone in the emergency department: A randomized controlled study. Ann Emerg Med 1999; 33:487.

13. Manning H, Schwartzstein R. Pathophysiology of dyspnea. N Engl J Med 1995; 333:1547.

14. Manthous C. Management of acute asthma. Res & Staff Phys 2000; 46:11.

15. Mihm F, Halperin B. Noninvasive detection of profound arterial desaturations using a pulse oximetry device. Anesthesiology 1985; 62:1.

16. Nicolaou D and Kelen G. Acid-base disorders. In: Tintinalli J et al. Emergency Medicine: A Comprehensive Study Guide. New York: McGraw-Hill, 2000.

17. Paape K, Fry W. Spontaneous pneumothorax. Chest 1994; 4:517.

18. Shapiro B, Cane R. Blood gas monitoring: Yesterday, today and tomorrow. Crit Care Med 1989; 17:966.

19. Sherter C, Hill D. Update on the treatment of asthma. Res & Staff Phys 2000; 46:5.

20. Stein P et al. Clinical characteristics of patients with acute pulmonary embolism. Am J Cardiol 1991; 68:1723.

21. Tapson V. Management of the critically ill patient with pulmonary embolism. J Crit Illness 2000; 15:S18.

22. Vines D et al. Current respiratory care, Part 1: Oxygen therapy, oximetry, bronchial hygiene. J Crit Illness 2000; 15:507.

23. Ward K. Pulse oximetry. In: Wolfson A, Paris P, eds. Diagnostic Testing in Emergency Medicine. Philadelphia: WB Saunders, 1996.

24. Yelderman M, New W. Evaluation of pulse oximetry. Anesthesiology 1983; 59:4.

4

CHAPTER 5

Vital Sign #4: Blood Pressure

Anatomy and Physiology

Blood pressure is highest in thick-walled arteries nearest the heart and lowest in veins which are thinner and away from the heart. A pressure of 100 mmHg in the brachial artery, close to the heart, means a force sufficient to push a column of Mercury up 100 millimeters in a manometer tube.

Regulation of Blood Pressure

Sympathetic fibers (efferents) from the vasomotor center in the medulla innervate arterioles. Arterioles regulate blood pressure. Normally they are in a state of partial constriction, or arteriolar tone. Several types of adrenergic receptors exist in the autonomic nervous system. Alpha, beta-1 and beta-2 are found on many organs. Alpha receptors are predominant in smooth muscle of arterioles. Stimulation of the sympathetic portion of the vasomotor center causes vasoconstriction and a rise in blood pressure; nonstimulation causes vasodilation and a fall in blood pressure. Factors affecting the vasomotor center and blood pressure are:

1. **Cardiac status.** In general, factors causing an increase and decrease in heart rate cause an increase and decrease in blood pressure. Tachycardia elevates blood pressure. Bradycardia or severe tachycardia lowers blood pressure because of decreased cardiac output.

2. **Baroreceptors** are nerve endings sensitive to pressure, or stretch (see Fig. 4.6, Chapter 4). The more important are located near chemoreceptors at the arch of the aorta and at the bifurcation of carotid arteries. Afferents run to the vasomotor center via the glossopharygeal and vagus nerves. An increase in blood pressure stimulates the vagal portion of the vasomotor center in the brainstem, the blood pressure decreases and the heart rate slows. Baroreceptor stimulation by massaging the carotid artery area on one side (**carotid massage**) in the emergence department may convert supraventricular tachycardia to a normal sinus rhythm. In hypotension, baroreceptors are not stretched and sympathetic output is increased, causing alpha and beta-1 stimulation, vasoconstriction and tachycardia.

3. Other factors may act on the vasomotor center to increase or decrease blood pressure. Emotions play a role: nerve fibers from the cortex synapse in centers in the hypothalamus, and stimulation may cause an increase or decrease in blood pressure. Stimulation of the lateral spinothalamic tract (pain and temperature pathway to the brain) causes a rise in blood pressure. When oxygen is low and carbon dioxide is high, as in hypotension, chemoreceptors are activated. Chemoreceptor afferents pass with baroreceptor fibers to the vasomotor center, assisting in elevating the blood pressure. An increase in carbon dioxide causes vasodilation of the vessels, particularly in the brain and skin. A decrease causes vasoconstriction.

Normal Blood Pressure

In the average adult, a pressure greater than 160/90 mmHg is considered high (hypertension) and a pressure less than 90/60 mmHg low (hypotension). Blood pressure may be measured directly by arterial cannulation, or indirectly using a sphygmomanometer. **Direct monitoring** is performed on critically ill patients: an angiocatheter is inserted into the radial artery and connected to a pressure transducer/monitor. With the exception of severe hypotension, **indirect monitoring** is usually within 10 mmHg of direct monitoring.

The mean pressure, or **mean arterial pressure (MAP)** is the average of pressures recorded over a period of time (not an average of systolic and diastolic pressures). One-third of the pulse pressure is added to the diastolic pressure. Example: what is the mean pressure of 140/80 mmHg? The pulse pressure is 40 mmHg. 1/3 of 40 = 13, and 13 + 80 = 93. Answer: the mean pressure is 93 mmHg.

Blood Pressure Devices

1. The **mercury sphygmomanometer** (sphygmo—Gr: pulse) is the usual apparatus for measuring blood pressure. It consists of an inflatable sleeve or cuff of various widths and lengths attached by tubing to a manometer tube containing Mercury. It measures the pressure required to move a column of Mercury up the tube a certain distance.
2. The **aneroid** (sphygmo) **manometer** (aneros—Gr: no liquid) uses a similar cuff to the above but air pressure elongates a metal bellows that transmits motion to the needle in a small round device that one may hang from the cuff. These are popular with physicians. However, many are inaccurate because of infrequent calibrations.
3. The **automatic electronic oscillometric instrument** (i.e., Dinamap) is widely used in hospitals today. The cuff is connected to a monitor which may also display oxygen saturation and cardiac rhythm.

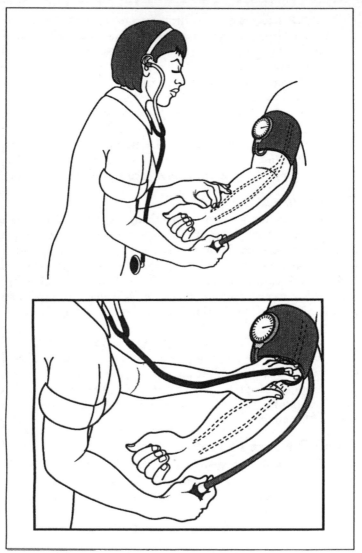

Fig. 5.1. Blood-pressure Reading.

A microprocessor senses the amplitude of arterial wall oscillations. As the cuff deflates, the first increase in amplitude is the **systolic** pressure; a quick decrease is the **diastolic** pressure. The **mean pressure** is the strongest amplitude. The devise is statistically more accurate than the mercury sphygomomanometer. Recent **portable devices** have arm, wrist or finger cuffs (the latter two are prone to error). A built-in algorithm calculates blood pressure based on arterial vibration patterns. Some portable monitors also have oxygen saturation capabilities (i.e., OscilloMate NIBP Monitor, CASE Medical Systems, Inc.)

Indirect Measurement of Blood Pressure

Blood pressure should ideally be assessed in both arms. The patient is positioned so that the elbow is flexed and perhaps supported on a pillow. The cuff of the mercury or aneroid manometer is placed so that the lower edge is about an inch above the antecubital fossa. Do not wrap the cuff over clothing or the reading will be falsely high. A 5 inch cuff is used in adults. The bladder of the cuff should encircle at least 2/3 of the arm. In an obese person, it is wise to use a large or leg cuff 8 inches wide or the reading will be 10-15 mmHg higher than the actual value. In infants and small children, a 1 1/2 inch cuff is used. Ages 2-5 years require a 3 inch cuff (see Chapter 7).

Palpate the radial artery and inflate the cuff until pulsations disappear. Note the pressure. Deflate the cuff, palpate and place the stethoscope on the **brachial artery** at heart level. An elevated position above the heart will give a falsely low reading; one below the heart will give a falsely high reading. Inflate the cuff 20-30 mmHg above where radial artery pulsations disappear, then lower the pressure slowly until sounds appear. These are the **sounds of Korotkoff** and are divided into several phases. Phase I is when they are first heard, and is the **systolic pressure**. Phase II corresponds to Phase I with a swishing sound, Phase III occurs when the sounds are loudest, Phase IV is the muffling of sounds and Phase V is when sounds are no longer heard.

Controversy exists over whether Phase IV or V is the true **diastolic pressure**. The point at which sounds are no longer heard is usually recorded. If the difference between the muffling and disappearance of sounds is less than 10 mmHg, it makes little difference which is recorded. If greater than 10, it is prudent to record both (i.e., 160/80/60). Sometimes sounds are heard to 0. Obviously a diastolic pressure of 0 is impossible, so the muffling of sounds is the diastolic pressure. This is seen in such conditions as hyperthyroidism and aortic regurgitation.

Occasionally, while deflating the cuff, sounds appear, then disappear, then reappear. This **auscultatory gap**, caused by diminished blood flow to the extremity, is seen in conditions such as arteriosclerotic disease, hypertension and aortic stenosis. Sounds are actually present, but inaudible. The gap has important consequences for the blood-pressure taker. If the cuff is only inflated to the gap, the systolic pressure will be falsely low. If the systolic pressure is noted correctly and the first disappearance of sounds is taken as the diastolic pressure, the diastolic will be falsely high. This is overcome by palpating the radial artery while the cuff is deflated, making sure sounds are auscultated for a sufficient amount of time. The gap is usually within 40 mmHg of the systolic pressure.

The opposite arm is used if an IV is running, and the affected arm is not used in trauma and in the postmastectomy and renal patient with an arteriovenous fistula. Lower extremity arteries are used when trauma, including burns, is present in both upper extremities. Each time blood pressure is taken the cuff should be completely deflated. Multiple attempts at blood pressure taking without deflating the cuff not only irritates the patient but produces a falsely elevated reading.

Blood-Pressure by Palpation

In some critically ill patients, it is occasionally not possible to auscultate a blood pressure. Palpation may be performed, similar to auscultation: the cuff is inflated to 20 mmHg above the level at which a palpable brachial pulse disappears and deflated until it appears. The result is an **estimated systolic pressure**.

Forearm and Leg Blood-Pressures

If the brachial artery is unavailable, the cuff may be placed around the forearm and the radial artery auscultated. The systolic pressure is 10 mmHg lower than the brachial. If the arm is unavailable, the thigh or leg may be used. The person lies on his stomach, or on his back with the knee flexed, and an 8 inch cuff is wrapped around the thigh. Inflate the cuff as the popliteal artery is auscultated (a difficult artery to palpate). The systolic pressure is 20 mmHg higher than the brachial systolic pressure. The diastolic is the same. The cuff may be wrapped around the leg just above the malleoli. Either the dorsal pedis or posterior tibial artery may be used. Systolic and diastolic values are the same as for the brachial artery (Fig. 5.2).

Doppler Stethoscope

The Doppler Stethoscope is a transducer with a high-frequency output that measures flow. A weak nonpalpable pulse may be auscultated, and systolic pressures as low as 30 mmHg may be detected (the diastolic cannot be measured). The cuff is placed around the arm and inflated, the Doppler is placed over the artery and the systolic pressure is noted when pulsatile sounds are first heard. Because it evaluates flow rather than pulsations, the Doppler is more sensitive than a regular stethoscope (see Fig. 3.22).

Approximation of Blood-Pressure from Pulse

In the past, it was alleged but never substantiated by invasive monitoring that a palpable radial, femoral and carotid pulse meant a systolic pressure of >80 mmHg, a carotid and femoral pulse represented a pressure between 70 and 80 mmHg and a palpable carotid pulse indicated a pressure between 60

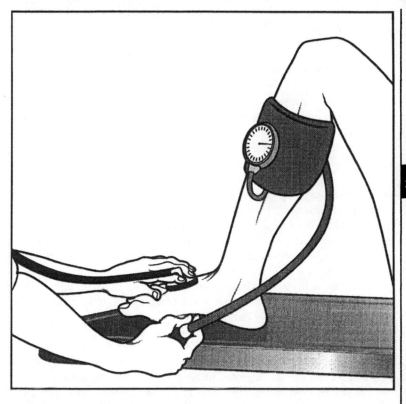

Fig. 5.2. Leg Blood-Pressure.

and 70 mmHg. Since invasive confirmation was never obtained and since pressures seemed to vary widely the concept faded.

Recently, a British anesthetist chose to challenge the 1985 ATLS guidelines showing this relationship, not realizing that the current version of the ATLS manual (1997) no longer includes it. He performed invasive monitoring on 30 patients with hypotension secondary to hypovolemic shock, grouping them as follows: group 1—radial, femoral and carotid pulses, group 2—femoral and carotid pulses, and group 3—carotid pulse only. In group 1 the high pressure was 88 mmHg with a mean of 72.5 mmHg, group 2 had a high of 78 mmHg with a mean of 66.4 mmHg, and group 3 had a high of 56 mmHg with a mean of 50 mmHg. A fourth group showed no palpable pulses, but had systolic pressures of 52, 54 and 76 mmHg.

The results indicate a lower than expected correlation, with a wide variation in predicted pressures from pulses. **In summary**, in this study a palpable radial pulse indicated a pressure between 53 and 88 mmHg, a femoral indicated a pressure between 48 and 78 mmHg, and a carotid indicated a pressure between 38 and 58 mmHg. The interesting facet of this investigation is that a substantial blood pressure existed in the absence of any palpable pulse.

Pulsus Paradoxus

As mentioned in Chapter 1, Kussmaul in 1873 noticed that in patients with pericardial effusion the pulse decreased and sometimes disappeared during inspiration. Later, after development of the blood-pressure cuff, it was found that a decrease in blood-pressure accompanied the weak pulse. In a normal person a slight waxing and waning of blood pressure exists during inspiration—up to 10 mmHg. Today a **pulsus paradoxus** is defined as an inspiratory fall in systolic blood-pressure greater than 10 mmHg. It is measured as follows: using normal blood-pressure protocol, the cuff is inflated above systolic pressure, then deflated to Korotkoff sounds during expiration. While the pressure is maintained the person inhales. The cuff pressure is lowered until Korotkoff sounds reappear. If this exceeds 10 mmHg, a paradoxical pulse is present. It is seen in conditions such as **cardiac tamponade**, severe asthma, COPD and heart failure. Its usefulness in a busy emergency setting is limited. Palpation of the radial or femoral pulse during inspiration is more practical, particularly if cardiac tamponade is suspected (see Chapter 3, Paradoxical Pulse).

Increased Pulse Pressure

The **pulse pressure** is the difference between systolic and diastolic pressures and reflects cardiac output. The systolic pressure rises in exercise, fever and conditions such as hyperthyroidism. Attention to the pulse pressure is important in head injuries, because a widening pulse pressure may indicate rising intracranial pressure from edema or bleeding (see Chapter 6). **Aortic regurgitation** may cause a widened pulse pressure. When the aortic valve is damaged, some aortic blood flows backwards at diastole, lowering the diastolic pressure. The left ventricle now pumps more blood during systole, raising the systolic pressure. **Treatment:** valve replacement.

In **patent ductus arteriosus**, the connection between the pulmonary artery and aorta remains open in the adult. In the fetus, the ductus arteriosus carries blood from the pulmonary artery to the aorta, bypassing the lungs, which are inactive. If the ductus fails to close at birth, blood is shunted from the high-pressure aorta back into the pulmonary artery. The result is high output heart failure (the left ventricle is pumping two or more times the normal cardiac output). A continuous "machinery" murmur is heard over the pulmonary area. **Treatment:** surgical correction.

Decreased Pulse Pressure

A narrowed pulse pressure is seen in conditions such as hypovolemic shock and cardiac tamponade. It is sometimes seen in heart failure from increased peripheral resistance and in severe aortic stenosis from decreased stroke volume.

High Blood Pressure (Hypertension)

Causes of Hypertension

The World Health Organization's criterion for a diagnosis of hypertension is a blood pressure of 160/95 mmHg or greater. Pressures between this and 140/90 are borderline. Most physicians treat diastolic pressures of 95 mmHg. Three readings are usually recorded over several days before treatment is begun. Factors such as anger, anxiety, stress and exercise may cause a transient increase in blood pressure so the person should rest for several minutes before the blood pressure is taken.

Primary Hypertension

In most cases (95%) the cause of hypertension is unknown (primary or "essential" hypertension). A genetic component appears to be present involving excessive sensitivity to dietary salt and an overstimulation of the sympathetic nervous system. The overstimulation of sympathetic activity causes arteriolar spasm which creates hypertrophy of the muscle layer of the arteriole, narrowing of the vessel, and an elevation of the blood pressure. The baroreceptor mechanism is reset to a higher level. The kidney appears to be involved. **Treatment:** weight reduction, salt restriction, and the discontinuance of alcohol and smoking. A stepped-care approach is often used: the patient is started on a diuretic, beta-blocker or calcium-channel blocker, and other agents are added as needed, such as one of the above or an angiotensin-converting enzyme (ACE) inhibitor.

Hypertensive Emergencies

Chronic uncontrolled or poorly controlled hypertension predisposes to a cerebrovascular event, such as hemorrhage, transient ischemic attack (TIA), thrombosis and hypertensive encephalopathy, as well as aortic dissection, cardiac disease (angina, myocardial infarction, congestive heart failure) and renal failure.

Hypertensive Encephalopathy

Hypertensive encephalopathy is seen when the blood pressure exceeds the limits of autoregulation by the blood-brain barrier. Blood enters brain tissue causing cerebral edema. The process develops over hours to days, with symptoms of headache, nausea, vomiting and an altered mental status ranging from confusion to lethargy. Visual changes, seizures and focal neurological changes may occur. The physical exam often reveals papilledema and/or retinal hemorrhages. Pressures may exceed 250/130 mmHg. **Treatment:** nitroprusside (Nipride) 0.5 ug/kg/min IV is a fast acting arterial and venous dilator. Labetalol (Normodyne) in 20mg IV increments may also be used.

Aortic Dissection

Aortic dissection is a tear of the thoracic aorta at the arch. Blood dissects through the tunica intima into the tunica media. The typical patient is an older hypertensive male with sudden onset of severe chest pain of a "tearing" quality, radiating to the back. A proximal dissection affecting the aortic valve and heart results in the diastolic murmur of aortic insufficiency, and possible pericardial effusion and tamponade. Involvement of the carotid arteries may result in signs of stroke. Paraplegia may be present if the vertebral and spinal arteries are involved. Pulse differences are aften present in the extremities. Tachycardia and signs of inadequate organ perfusion such as clammy skin and delayed capillary refill may be present. A chest x-ray often shows a widened mediastinum. **Treatment:** a thoracic surgeon should be immediately notified. A beta-blocker such as labetalol (Normodyne) 20 mg IV is administered, followed by nitroprusside (Nipride), beginning at 0.5 µg/kg/min, to maintain the systolic pressure at about 120 mmHg.

Hypertension and Cerebrovascular Accident

In hypertension accompanying a cerebrovascular accident, it is sometimes difficult to determine whether hypertension is the cause or the result of the problem. Increased blood pressure is frequently a response to stroke, although the patient with a **thrombotic** or **embolic stroke** usually has only a small elevation. As with an ischemic stroke, hypertension may contribute to an **intracerebral hemorrhage**, or be the result of it. **Subarachnoid hemorrhage** is seen in a younger population and is the result of a ruptured cerebral aneurysm or a bleeding arteriovenous malformation. **Treatment:** blood pressure management is seldom required for an ischemic stroke. If the diastolic rises over 130 mmHg, increments of labetolol (Normodyne) 10 mg IV every 20 minutes may be given to reduce the diastolic to slightly above prestroke levels for both hemorrhagic and ischemic strokes (for further information, as well as the treatment for **increased intracranial pressure**, see Chapter 6 and Fig. 6.3).

Hypertension and Cardiac Emergency

As with an acute cerebrovascular accident, it is often difficult to determine whether the hypertension caused the angina, myocardial infarction or pulmonary edema, or was the result of an alteration in left ventricular performance secondary to increased afterload that raised the blood pressure.

Treatment: nitroglycerine, a dilator of coronary arteries, is begun at 10 μg/min IV. Further control may be needed by nitroprusside 0.5 μg/kg/min. Treatment for acute pulmonary edema is discussed in Chapter 4 (see Fig. 4.12).

Secondary Hypertension

In 5% of cases the cause of hypertension is known (secondary hypertension). An important cause is hypertension related to pregnancy, which includes gestational hypertension, **pre-eclampsia** and eclampsia. Gestational hypertension is a blood pressure of 140/90 mmHg or greater after the 20th week of pregnancy. Pre-eclampsia, often seen in primagravidas, consists of hypertension, proteinuria and sometimes edema (eclampsia is pre-eclampsia with seizures, which may occur if pre-eclampsia is untreated). Headache and visual disturbances are common. Treatment for pre-eclampsia is hospitalization, fetal monitoring, and intravenous magnesium sulfate for seizure control. For a diastolic pressure of 110 mmHg or greater, hydralazine (Apresoline) 5 mg IV may be given. After the 36th week, induction of labor is the preferred treatment.

Renovascular hypertension is a common cause of secondary hypertension. **Treatment:** in renal failure, nitroprusside is begun at 0.5 μg/kg/min IV with the goal of maintaining the diastolic pressure at about 100 mmHg. Dialysis may be required.

Other causes of secondary hypertension are hypersecretion of steroid hormones from the adrenal cortex in **Cushing's disease** and **Conn's syndrome** (primary aldosteronism), and **pheochromocytoma**, which is an adrenaline-secreting tumor of the adrenal medulla causing episodic spells of headache, sweating and heart palpitations from surges of epinephrine and norepinephrine. **Treatment:** if the cause of Cushing's disease, Conn's syndrome or pheochromocytoma is a tumor, removal is the therapy.

A rare cause of episodic hypertension in a young person is **coarctation of the aorta**, a congenital narrowing of the thoracic aorta near the left subclavian artery causing high pressures in the upper body and low in the lower. Blood pressure must be determined in the arms and legs to detect the condition. **Treatment:** aortic resection or balloon angioplasty.

Drugs or drug withdrawal may cause a hypertensive crisis. Cocaine and amphetamines stimulate the adrenergic nervous system. **Treatment:** most of the time the cocaine abuser with hypertension responds to diazepam (Valium) 5 mg IV or lorazepam (Ativan) 2 mg IV. In severe cases nitroprusside is added.

Occasionally a person taking one of the **monoamine oxidase inhibitor antidepressants** indulges in Chianti wine, beer, cheese, or pickled herring containing the amino acid tyramine. Tyramine releases norepinephrine from sympathetic nerve endings normally inactivated by monoamine oxidase. Suppression of the mechanism causes excessive norepinephrine release and a hypertensive crisis. **Treatment:** nitroprusside 0.5 μg/kg/min.

Withdrawal from alcohol, opiates, or noncompliance with antihypertensive drugs such as clonidine (Catapres), may also cause a hypertensive event. **Treatment:** the hypertension of alcohol withdrawal usually responds to diazepam (Valium) 5-10 mg IV or lorazepam (Ativan) 2 to 4 mg IV followed by diazepam 10 mg or chlordiazepoxide (Librium) 50 mg PO every 6 hours. The abrupt cessation of antihypertensive drugs such as clonidine may cause rebound hypertension. Treatment consists of reinstitution of the drug and tapering.

Low Blood Pressure (Hypotension)

Shock represents inadequate circulatory perfusion to meet metabolic demands. A significant sign is **hypotension**. Types of shock are hypovolemic (i.e., hemorrhage, dehydration), cardiogenic (myocardial infarction, cardiac tamponade), septic, neurogenic and anaphylactic. The systolic pressure is usually low (below 70 mmHg; however, a hypertensive patient may be in shock at a pressure of 120/80). The body's automatic responses are activated. The prototype is seen in hypovolemic shock:

1. The **baroreceptor** mechanism is stimulated.
2. Arterioles in skin, muscles, kidneys and the GI tract containing alpha receptors constrict. Blood is shunted to vital organs such as the heart and brain. The result is cool, clammy skin, decreased urinary output and fluid retention.
3. Oxygen is low, carbon dioxide is high and **chemoreceptors** are stimulated, assisting in elevated blood pressure.
4. The sympathetic nervous system is stimulated and **norepinephrine** is released from sympathetic nerve endings.
5. Stimulation of the sympathetic system causes release of **epinephrine** and norepinephrine from the adrenal medulla, resulting in further vasoconstriction and tachycardia.
6. Hypotension causes **vasopressin** (antidiuretic hormone, ADH) release from the posterior pituitary gland. In addition to its antidiuretic effect on the kidney, ADH is a vasoconstrictor.
7. The kidney secretes the enzyme renin, which acts in the lungs form **angiotensin II**, a potent vasoconstrictor.

Baroreceptor and chemoreceptor control mechanisms occur in seconds. Hormonal and kidney mechanisms require several minutes.

Hypovolemic Shock

Common causes of hypovolemic shock are hemorrhage from trauma and gastrointestnal bleeding (i.e., ulcers). A less common cause is dehydration from vomiting, diarrhea or low fluid intake. A 15% blood/fluid loss causes tachycardia. A 15-30% loss causes tachycardia, tachypnea, decreased pulse pressure and prolonged capillary refill. Only when the loss is about 30-40%

does the systolic pressure begin to drop. The skin is cool and clammy, accompanied by restlessness and anxiety. **Treatment:**
1. External hemorrhage is controlled by pressure;
2. Two large bore IV's are placed and 2 liters (peds: 20ml/kg x 3) of normal saline or lactated Ringer's solution is infused wide open to maintain a urine output of 30 cc/h (peds: 1 cc/kg/h);
3. A loss of 30% of blood volume requires the administration of blood (type-specific packed cells or O-negative in an emergency);
4. Possible traumatic abdominal hemorrhage requires a diagnostic peritoneal lavage and
5. A surgical consultation is required.

A normal heart-rate or bradycardia rather than tachycardia is sometimes seen in hemorrhagic shock (5-50%) of cases). It is called **paradoxical** or **relative bradycardia**, although the heart rate is normal (60-100) in most cases (the median rate is 80, although a few cases are <60). A more precise term would be **hemorrhagic non-tachycardia**. Originally thought to be a vagal response to blood in the peritoneal cavity (i.e., abdominal trauma, splenic rupture, bleeding ovarian cyst, ruptured ectopic pregnancy), it was later discovered in thoracic and extremity trauma as well.

The reason for the response is the following: at a loss of about 15% body fluid/blood, sympathetic activity is increased and vasoconstriction and tachycardia occur. As blood loss approaches 30% and the systolic pressure decreases the left ventricle is now contracting around a reduced volume. This triggers stimulation of unmyelinated afferent vagal fibers in the left ventricle and bradycardia occurs. This reflex (vago-vagal reflex) prevents further sympathetic stimulation and reduction of left ventricular volume, preserving organ perfusion. Evidence: the efferent response is abolished by atropine. As blood volume and pressure decline further, the baroreceptor response overrides the reflex and tachycardia resumes, continuing until terminal bradycardia and cardiac arrest occur.

The reflex does not seem to be rate-dependent ,and it is not consistently seen. Current thinking is that the reflex is often overridden by the sympathetic/baroreceptor response. It is not present in hypovolemia from dehydration.

In summary, an increase in heart-rate is a useful parameter for the assessment of bleeding, but its absence does not rule out severe hemorrhage. Profound shock may occur with a normal heart rate or bradycardia. It is thus an unreliable assessment tool. Hypotension and an alteration in behavior are more reliable signs. **Treatment:** fluid/blood resuscitation at all stages, as noted previously.

Orthostatic Vital Signs

Confusion exists with this topic, sometimes for conceptual reasons but also because of ambiguous terminology. Orthostatic vital signs are **heart-rate** and

	Class I	Class II	Class III	Class IV
Blood Loss (mL)	Up to 750	750-1500	1500-2000	>2000
Blood Loss (% Blood Volume)	Up to 15%	15%-30%	30%-40%	>40%
Pulse Rate	<100	>100	>120 (<100)	>140
Blood Pressure	Normal	Normal	Decreased	Decreased
Pulse Pressure (mm Hg)	Normal or Increased	Decreased	Decreased	Decreased
Respiratory Rate	14-20	20-30	30-40	>35
Urine Output (mL/hr)	>30	20-30	5-15	Negligible
CNS/Mental Status	Slightly anxious	Mildly anxious	Anxious, confused	Confused, lethargic
Fluid Replacement (3:1 Rule)	Crystalloid	Crystalloid	Crystalloid and blood	Crystalloid and blood

Fig. 5.3. Evaluation of Fluid/blood Loss. Modified with permission from American College of Surgeons: Committee on Trauma, Advanced Trauma Life Support for Doctors, Student Course Manual, 6th ed., p. 98. ©1997 American College of Surgeons.

blood-pressure. The word "orthostatic" means assuming an erect position. When a person stands upright the heart-rate increases slightly (about 10 beats per minute), the systolic pressure decreases slightly and the diastolic increases slightly (compensatory baroreceptor activity). As the body loses about a liter of fluid, gravity begins to have an effect. The first sign of hypovolemia is an increase in heart-rate, followed by a slight decrease in pulse pressure.

Orthostatic tachycardia is the correct but rarely used term for "positively orthostatic". In a patient with blood loss or dehydration, an increase in pulse-rate of 30 beats per minute on standing represents a blood/fluid loss of about a liter. Figure 5.3 indicates that this is about a 20% loss. When about 35% of the blood volume is lost (about 2 liters in the average adult), the systolic pressure begins to drop (orthostatic tachycardia and hypotension).

Being "positively orthostatic" may also mean having **orthostatic (postural) hypotension**, or a drop in blood-pressure (>20/10 mmHg) on standing **without** an increase in heart-rate. This is usually not a sign of hypovolemia but occurs in patients on beta-blockers, alpha-blockers, calcium-channel blockers, nitrates, phenothiazines, with alcohol ingestion (impairment of vasoconstriction) and in the rare person with autonomic dysfunction. It may be seen with hypovolemia in the occasional elderly patient with a weak baroreceptor response, and as mentioned in the previous section, with blood loss resulting in **paradoxical bradycardia**. Paradoxical bradycardia is not seen in hypovolemia from dehydration.

Procedure:

1. *Blood pressure and pulse are taken after patient supine for 3 minutes.*

2. *After standing (or sitting if standing not possible) for 1 minute, pulse and blood pressure retaken. If patient becomes faint or dizzy, the patient lies down and pulse and blood pressure retaken.*

Results:

1. *If pulse increases 30 beats or more after standing (or sitting) for one minute, or if pulse increases and near-syncope develops, the patient has* **orthostatic tachycardia,** *from hypovolemia.*

2. *If pulse increases 30 beats or more and blood pressure decreases > 20/10 mmHg after standing (or sitting) for 1 minute, the patient has* **orthostatic tachycardia and hypotension,** *from severe hypovolemia.*

3. *If pulse unchanged but blood pressure decreases after standing (or sitting) for 1 minute, or near-syncope develops, the patient has* **orthostatic hypotension,** *most probably from meds.*

Fig. 5.4. Orthostatic Vital Signs

Orthostatic vitals are recorded as follows: the patient lies for three minutes and the blood-pressure and heart-rate are recorded. He then stands (or sits up) for one minute and they are retaken. The patient has orthostatic tachycardia (positively orthostatic) if the heart-rate increases by 30 beats per minute, or if he becomes dizzy or light-headed with a lesser increase. Orthostatic hypotension is present if only the blood pressure decreases >20/10 mmHg. Both conditions are considered "positively orthostatic".

Orthostatics are often reported by using a stick-figure lying, sitting or standing, with the appropriate heart-rate and blood-pressure readings adjacent to the figure.

An older term, the **tilt test**, is occasionally encountered in the literature. The test was originally designed with the patient lying, then the patient was tilted to a sitting position and the heart-rate and blood-pressure retested. The test was positive if the pulse rate increased 15 to 20 beats, or the systolic blood pressure decreased 15 to 20 mmHg. The test is no longer considered valid (orthostatics from lying to sitting will not detect a 1000 cc blood/fluid loss). However, if the heart-rate increases 30 or more beats per minute or the person becomes dizzy or light-headed from lying to sitting, this is a positive test. Results were written as "tilt positive" or "tilt negative". When practitioners today speak of a tilt test, they mean (hopefully) current orthostatics.

The elderly and and occasionally others do not always follow the rules. Cardiac syncope and poor autonomic functioning may result in false positives or negatives. An entity in some elderly, **postural orthostatic tachycardia**

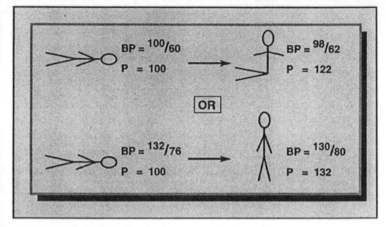

Fig. 5.5. Orthostatic Figures

syndrome, is a fall in blood pressure, tachycardia, near-syncope and symptoms of a transient ischemic attack (TIA), thought to be caused by autonomic dysfunctioning and not hypovolemia.

The accuracy of orthostatic vital signs is frequently challenged. A recent study showed that some normal subjects had a heart-rate increase from 5 to 39 beats per minute (with a mean of 17) from supine to standing. Both systolic and diastolic pressures rose. The cause was related to baroreceptor stimulation with both a-adrenergic and b-adrenergic effects. The conclusion was that a wider than previously thought variability exists in the physiologic response to standing.

To conclude, orthostatics are of value if the procedure is accurately performed and accompanied by a careful history. Excluding other factors, lightheadedness or dizziness on standing or sitting upright is a positive test regardless of the heart-rate. Patients are located within a bell-curve. At the edges are rare false positives and negatives. **Treatment:** Fluid/blood resuscitation as previously described.

Capillary Refill

The capillary refill is the time it takes for blanching of the skin to return to normal when the nailbed or hypothenar emminence is quickly squeezed. It is a simple measurement of blood/fluid loss. The upper limit of normal in males is 2 seconds; in females 2.9 seconds. A delay greater than this implies a fluid deficit of about 100 cc/kg. The test has been challenged. Recent studies indicate that its value as an isolated test for mild-to-moderate hypovolemia is minimal and even misleading. However, if orthostatics are abnormal, the validity increases. The test is more sensitive in children (see Chapter 7). **Treatment:** Fluid/blood resuscitation as previously described.

Abdominal Aortic Aneurysm (AAA)

A rare but potentially catastrophic cause of hypovolemic shock is a leaking abdominal aortic aneurysm. Because of a weakened tunica media from atherosclerosis, the abdominal aorta may slowly balloon out. A common presentation is an older male with sudden onset of severe lower abdominal and back pain, sometimes radiating to the groin and accompanied by hypotension. This represents a leaking aneurysm. Lower extremity pulses may or may not be weak compared to upper ones. The physical exam reveals a tender abdominal pulsatile mass. **Treatment:** a call is made to the surgeon for immediate surgery, while 2 IV's are started and blood is typed and crossed for 10 units.

Cardiogenic Shock

Cardiogenic shock is pump failure, usually caused by an acute myocardial infarction (involving about 40% of the myocardium), but occasionally from cardiomyopathies, drugs, toxins, pulmonary embolism, cardiac tamponade and some arrhythmias.

Cardiogenic Shock from Myocardial Infarction

The patient may migrate into a shock state from a heart attack or present in shock. The main symptom is chest pain, although in the elderly, and occasionally in others it is absent. In shock, the skin is cool and moist. Neck veins are distended. Auscultation of the heart may reveal an S-3 gallop, a new murmur, tachycardia or bradycardia. The EKG usually shows a pattern of acute injury. Rales may be present. A chest x-ray may show pulmonary edema. Serum markers (i.e., CK-MB and troponin) are usually positive. **Treatment for myocardial infarction:**

1. The ABCs of resuscitation are followed (i.e., high flow oxygen, pulse oximetry, intubation, IV access);
2. Aspirin 160 mg is chewed and swallowed;
3. Nitroglycerine 10 μg/min is given for pain, and also reduces preload and afterload. If pain does not resolve, morphine sulfate 2-5 mg is used;
4. For a systolic pressure less than 100 mmHg, a vasopressor is administered (see next section) (however, if a right ventricular infarction is present, a fluid challenge of normal saline is used);
5. Heparin 80 units/kg IV bolus and 18μ/kg/hr is administered;
6. A beta-blocker such as metoprolol (Lopressor) 5 mg q for 5 min is given for three doses (assuming no bradycardia or hypotension) and
7. A thrombolytic agent such as alteplase (t-PA) (Activase) 100 mg over 1.5 h or tenecteplase (TNKase) 40 mg over 5 seconds is administered.

Treatment for cardiogenic shock: the patient in shock (systolic BP<90 mmHg, pulmonary edema) should be transferred as soon as posible to a facility with the capability for intra-aortic balloon pump placement and **percutaneous transluminal coronary angioplasty (PTCA)**. A balloon-tipped catheter is maneuvered into the blocked coronary artery; the balloon is inflated, dilating the narrowed artery and disrupting the atheromatous plaque. A stent may be placed.

If that is not possible, thrombolytic therapy is begun (see above section). For a systolic BP of 90-100 mmHg, accompanied by pulmonary edema, furosemide (Lasix) 80 mg and dobutamine 5 µg/kg/min IV are administered. If the systolic BP is 70-90 mmHg, dopamine is begun at 5 µg/kg/min. Norepinephrine 2 µg/min is used for a systolic BP of <70 mmHg (see Fig. 4.12).

Shock from Cardiac Tamponade/Tension Pneumothorax

Other causes of cardiogenic shock are **cardiac (pericardial) tamponade** and **tension pneumothorax**. In these cases, the heart is compressed—in tension pneumothorax by air in the pleural cavity pressing against the heart, and in cardiac tamponade by bleeding into the the pericardial sac. Commonly, cardiac tamponade is seen after a stab wound to the heart (which usually nicks a vessel), but occasionally after blunt trauma. The condition may also result from accumulated fluid secondary to metastatic disease (malignant pericardial effusion). **Pulsus paradoxus** may be present (see earlier section) A familiar scenario is a young male seen in a trauma center for a stab wound to the anterior chest who wants to leave, the wound being barely visible. He is later found slumped on the cart, barely conscious, with a faint pulse, decreased blood pressure and jugular venous distension (Beck's triad). **Treatment:** treatment for tension pneumothorax is needle decompression, followed by chest tube placement, discussed in Chapter 4.Treatment for cardiac tamponade is intravenous fluid infusion and immediate pericardiocentesis.

Septic Shock

Bacteremia is an invasion of the bloodstream by infectious micro-organisms. Severe bacteremia is **sepsis**. Release of toxins by microorganisms (i.e., gram negative bacteria) causes host macrophages to produce defense substances such as kinins, cytokines, complement and prostaglandins causing vasodilation, increased capillary permeability, leaking of plasma into tissues and a fall in blood pressure (**septic shock**). Disordered temperature regulation, myocardial depression and multiple organ system failures occur. Complications are disseminated intravascular coagulation (DIC) and adult respiratory distress syndrome (ARDS). Frequent sites are the urinary tract, GI tract and lungs. Extremes of age are particularly susceptible, as are burn victims, diabetics, cancer patients and those having undergone recent invasive procedures.

The common septic patient is an elderly person with a urinary tract infection that has progressed to a systemic problem. The patient is warm and flushed, mildly agitated, and the temperature is increased ("warm shock"). This often progresses to "cold shock" from peripheral vasoconstriction and hypotension. Obtundation is often present, accompanied by hyperventilation because of metabolic acidosis. A widened pulse pressure is frequently present. **Treatment:**

1. The ABCs of resuscitation are followed;
2. Two large bore IV's are placed and Ringers lactate solution is infused wide open to maintain a urine output of 30 cc/h (peds: 1 cc/kg/h);
3. If the blood pressure does not improve with fluid, a pressor such as dopamine 5 ug/kg/min is added, and
4. An intravenous antibiotic relevant to the source of infection is started, such as IV ceftriaxone 1 gm + gentamycin 1.5 mg/kg (peds: ceftriaxone 50 mg/kg).

Neurogenic Shock

Neurogenic shock results from trauma to the spinal cord. Muscle flaccidity and loss of reflexes below the injury (**spinal shock**) is followed by mild hypotension and bradycardia (**neurogenic shock**) from damage to descending sympathetic pathways in the cord. **Treatment** consists of:

1. The ABCs of resuscitation;
2. Proper spinal immobilization;
3. Intravenous normal saline to maintain a systolic pressure above 70 mmHg;
4. If unable to maintain the blood pressure, dopamine 5 ug/kg/min is added;
5. Bradycardia may be treated with atropine 1 mg IV every 5 minutes to a total dose of 3 mg or a pacemaker if needed;
6. Methylprednisolone 30 mg/kg over 15 min is administered, followed by an infusion of 5.4 mg/kg per hour (controversial) and
7. The patient is transferred to an appropriate spinal-cord facility.

Anaphylactic Shock

Anaphylaxis, or anaphylactic shock, is a severe allergic reaction. Common allergens are penicillin and bee/wasp venom. Mediators (i.e., histamine, leukotrienes, prostaglandins) are released from mast cells. Flushing and an itchy skin, a red rash (urticaria), shortness of breath and hypotension are present. Sometimes swelling of the tongue, pharynx and larynx (laryngeal edema) and wheezing (bronchospasm) occur. **Treatment:**

1. Immediate intravenous normal saline is begun to raise the blood pressure;
2. High flow oxygen by mask is monitored by pulse oximetry;
3. Epinephrine 0.4 cc subq (peds: 0.01 cc/kg) (1:1000), or for severe shock 1 to 10 cc (1:10,000) slow IV push;
4. An antihistamine such as diphenhydramine 50 mg IV (peds: 1 mg/kg) is given;
5. Methylprednisolone 125 mg IV (peds: 1 mg/kg)is administered;
6. A histamine receptor blocker such as ranitidine 50 mg IV (peds: 0.5 mg/kg) is given over 5 min, and
7. A nebulized beta-agonist (i.e., albuterol) is administered for bronchospasm.

Other

A low pressure is sometimes seen in myxedema, adrenal crisis, DKA, renal failure, as well as with the use of drugs such as beta-blockers, calcium-channel blockers, diuretics, opioids and sedative hypnotics.

Special Cases

Hypotension in Pregnancy

Cardiac output increases by 40% at the end of the first trimester. The heart rate increases 10-15 beats per minute. Systolic and diastolic pressures decrease about 10 mmHg in the second trimester. In late pregnancy, the supine position compresses the inferior vena cava causing a further decrease in blood pressure. The left lateral position relieves this effect.

Vaginal bleeding and hypovolemic shock in early pregnancy may occur from spontaneous abortion or ectopic pregnancy, and in late pregnancy from abruptio placentae or placenta previa. A postpartum hemorrhage may also occur. If a spontaneous abortion becomes inevitable with heavy vaginal bleeding, treatment is fluid resuscitation with normal saline, followed by dilatation and curettage.

An unstable **ectopic pregnancy**, indicated by abdominal pain, vaginal bleeding, hypotension and a positive pregnancy test, is diagnosed by culdocentesis (the stable patient is diagnosed by ultrasound). **Treatment:** 2 IV lines are started, packed red cells are given for hemodynamic instability, and the patient is prepared for surgery.

Abruptio placentae, or premature separation of the placenta in late pregnancy, is indicated by abdominal pain and vaginal bleeding. Treatment includes fetal monitoring, IV normal saline and packed red cells as needed.

Placenta previa is painless bleeding in late pregnancy. Because the placenta lies over the cervical os, a cesarean section may be required. **Treat-**

ment: fetal monitoring, IV normal saline and packed red cells are administered as needed.

Postpartum hemorrhage results from uterine atony and sometimes from retained fetal tissue. **Treatment** consists of IV normal saline, packed red cells as needed and oxytocin or methylergonovine 0.2 mg IM to contract the uterus and control bleeding. Curettage is performed to remove retained tissue.

The Dialysis Patient

Hypotension during or after dialysis is frequently seen because of fluid loss. Treatment consists of placing the patient in the Trendelenburg position and giving a fluid challenge of 200 cc of normal saline.

Practical Points

- First, the ABC's of resuscitation are followed (see Chapter 8).
- When taking the blood pressure, if the muffling of diastolic pressure is greater than 10 mmHg, two values should be noted: the beginning and disappearance of muffling.
  Example:
    BP 210/110/80 right arm, supine
- Orthostatic tachycardia is an increase in heart rate from lying to standing. If the heart rate increases 30 beats and/or neurologic changes occur (i.e., dizziness), hypovolemia is present.
  Example #1:
    Pulse 101, BP 130/90, right arm lying
    Pulse 132, BP 134/94, right arm standing
  Example #2:
    Pulse 106, BP 142/94, right arm lying
    Pulse 124, standing dizzy—terminated
- If the blood pressure lowers from lying to standing without a heart rate increase, this is orthostatic hypotension.
  Example:
    Pulse 76, BP 116/60, left arm lying
    Pulse 78, BP 100/56, left arm standing
- In an acutely ill or trauma patient, always take BP in both arms. The pressure in one may be compromised.

References

1. American Heart Association and the International Liaison Committee on Resuscitation (ILCOR). Guidelines 2000 for cardiopulmonary resuscitation and emergency cardiovascular care. Baltimore: Lippincott, Williams & Wilkins, 2000.
2. Bakris G. Severe hypertension in a young patient. Hosp Pract 1993; 28:10.
3. Barach P. Pulsus paradoxus. Hosp Phys 2000; 36:49.
4. Baraff L, Schriger D. Orthostatic vital signs: Variation with age, specificity, and sensitivity in detecting a 450 ml blood loss. Am J Emerg Med 1992; 10:2.

5. Bruce C et al. The effect of cocaine on the physiologic response to hemorrhagic shock. Surgery 1993; 114:429.
6. Califf R, Bengtson J. Cardiogenic shock. N Engl J Med 1994; 330:24.
7. Chen H et al. Enhancement of vagal restraint on systemic blood pressure during hemorrhage. Am J Physiol 1978; 234:192.
8. Cohn L. Aortic dissection: New aspects of diagnosis and treatment. Hosp Pract 1994; 29:3.
9. Craver J, Connolly M. The percutaneous intraaortic balloon pump and ventricular assist devices. In: Schlant R, Alexander R. The Heart. New York: McGraw-Hill, 1994.
10. Deakin C, Low J. Accuracy of the advanced trauma life support guidelines for predicting systolic blood pressure using carotid, femoral, and radial pulses. BMJ 2000; 321:673.
11. Demetriades D et al. Relative bradycardia in patients with traumatic hypotension. J Trauma 1998; 45:534.
12. Enselberg C. Measurement of diastolic blood pressure by palpation. N Eng J Med 1961; 265:6.
13. Evans R et al. Does the hemodynamic response to acute central hypovolaemia depend on the rate of fall of cardiac output? Clin Exp Pharmacol Physiol 1992; 19:657.
14. Grubb et al. The postural orthostatic tachycardia syndrome: A neurocardiogenic variant identified during head-up tilt table testing. Pacing Clin Electrophysiol 1997; 20:2205.
15. Hals G, Carleton S. Pericardial disease and tamponade. Em Med Rep 1996; 17:161.
16. Hals G, Crump T. The pregnant patient: Guidelines for management of common life-threatening medical disorders in the emergency department. Em Med Rep 2000; 21:53.
17. Hartmann A et al. Measurement of blood pressure in the brachial and posterior tibial arteries using the Doppler method. J Ped 1973; 82:3.
18. Jackson R. Cardiogenic shock. In: Tintinalli J et al. Emergency Medicine: A Comprehensive Study Guide. New York: McGraw-Hill, 2000.
19. Jacob G et al. The neuropathic postural tachycardia syndrome. N Engl J Med 2000; 343:1008.
20. Jensen K. Heart and endocrine changes during central hypovolemia in man. Danish Medical Bulletin 1991; 38:443.
21. Kaplan N. Establishing control of refractory hypertension. Hosp Pract 1994; 29:5.
22. Knopp R et al. Use of the tilt test in measuring acute blood loss. Ann Emerg Med 1980; 9:29.
23. Koziol-McLain J et al. Orthostatic vital signs in emergency department patients. Ann Emerg Med 1991; 20:6.
24. Kumar A et al. Hypertensive crisis. J Emerg Med 2000; 19:369.
25. Kussmaul A. Ueber schwielige Mediastino-Pericarditis und den paradoxen Puls. Berl Klin Wochenschr 1873; 10:37-39.
26. Lewinter J et al. Vital sign measurement procedures. In: RobeRTS J, Hedges J, eds. Clinical Procedures in Emergency Medicine. Philadelphia: WB Saunders, 1998.
27. Malhotra A, Townsend R. Clinical significance of systolic and pulse pressure. Em Med 2000; 32:52.
28. Mansoor G, White W. Ambulatory blood pressure monitoring: A clinically relevant tool for the diagnosis and management of hypertension. Res & Staff Phys 1999; 45:10.
29. Mansoor G, White W. Usefulness of home blood pressure monitoring in clinical practice. Res & Staff Phys 2000; 46:21.
30. McGregor M. Pulsus paradoxus. N Engl J Med 1979; 301:480.

31. Pahwa R, Dellinger R. Cerebral hypoperfusion with "normal" blood pressure. J Crit Illness 2000; 15:567.

32. Park M, Guntheroth W. Direct blood pressure measurement in brachial and femoral arteries in children. Circul 1970; 41:231.

33. Pascarelli E, Bertrand C. Comparison of blood pressures in the arms and legs. N Eng J Med 1964; 270:14.

34. Ram C. Secondary hypertension: workup and correction. Hosp Pract 1994; 29:4.

35. Rigolin V et al. Update on aortic dissection. Emerg Med 1993; 25:13.

36. Rodbard S, Margolis J. The auscultatory gap in arteriosclerotic heart disease. Circ 15:June, 1957.

37. Sander-Jensen K et al. Vagal slowing of the heart during haemorrhage: Observations from 20 consecutive hypotensive patients. Brit Med J 1986; 292:364.

38. Schmidt R. Shock. In: Markovchick V, Pons P. Emergency Medicine Secrets. Philadelphia: Hanley & Belfus, 1999.

39. Schriger D et al. Capillary refill—Is it a useful predictor of hypovolemic states? Ann Emerg Med 1991; 20:6.

40. Secher N et al. Bradycardia during reversible hypovolaemic shock: Associated neural reflex mechanisms and clinical implications. Clin Exp Pharmacol Physiol 1992; 19:773.

41. Shabetai R et al. Pulsus paradoxus. J Clin Investig 1965; 44:11.

42. Shabetai R et al. The hemodynamics of cardiac tamponade and constrictive pericarditis. Am J Cardiol 1970; 26:480.

43. Thompson D et al. Relative bradycardia in patients with isolated penetrating abdominal trauma and isolated extremity trauma. Ann Em Med 1990; 19:268.

44. Vayer J et al. Absence of a tachycardic response to shock in penetrating intraperitoneal injury. Ann Em Med 1988; 17:227.

45. Witting M et al. Defining the positive tilt test: A study of healthy adults with moderate acute blood loss. Ann Emerg Med 1994; 23:6.

46. Wo C et al. Unreliability of blood pressure and heart rate to evaluate cardiac output in emergency resuscitation and critical illness. Crit Care Med 1993; 21:2.

47. Wynn R. Obstetrics and Gynecology. Philadelphia: Lea & Febiger, 1988.

48. Yutaka I et al. Clinical evaluation of semiautomatic and automatic devices for home blood pressure measurement: Comparison between cuff-oscillometric and microphone methods. J Hypertension 1989; 7:983.

CHAPTER 6

Vital Sign #5: Level of Consciousness

The Glasgow Coma Scale, as well as AVPU (**A**lert, responds to **V**erbal stimuli, responds to **P**ainful stimuli, **U**nresponsive) have been rapid neurological assessment tools for prehospital and hospital personnel for many years, and to that extent level of consciousness has been a vital sign for over three decades.

The brain is quite sensitive to body changes, and an alteration in mental status often precedes abnormalities in other vital signs. Altered mentation may range from bizarre behavior and confusion to **coma**, a state of unresponsiveness from which the patient cannot be aroused. **Psychotic behavior** is usually lucid and not confused. The person with **delirium** is confused, and the onset is fairly sudden. Common causes of delirium are hypoglycemia in the diabetic, drugs, alcohol, inappropriate meds in the elderly, infection, withdrawal syndromes and hypoxemia. Recently, the serotonin syndrome, a complication of new antidepressant drugs, may cause agitation and confusion from increased central serotonin neurotransmission. Delirium is reversible. Untreated, it may progress to coma. **Dementia**, on the other hand, is a gradual loss of mental capacity, is primarily a disorder of the elderly and is rarely reversible. Alzheimer patients make up 70% of dementia cases, with multi-infarct dementia accounting for 15-20% of others. The three conditions may co-exist, particularly in the elderly. Overlaps and misinterpretations are frequent. Psychotic depression, for example, may be misdiagnosed as dementia. On the other hand, delirium from medicines may be misdiagnosed as depression. Therefore, it is important to rule out an organic cause for altered mentation before classifying behavior as a functional disorder.

Common causes of an altered level of consciousness are alcohol and drug abuse (30%), hypoglycemia from insulin reactions (30%) and stroke (30%). In some centers the percentages vary (i.e., trauma centers, hospitals that see many elderly).

Anatomy and Physiology

The cerebral hempispheres are two large areas of **white matter** surrounded by an outer thin mantel of nerve cell bodies, the **grey matter**, or **cortex**. Association areas constitute the majority of the cortex and are involved with higher learning or intelligence. Both cerebral hemispheres must be affected for an altered level of consciousness to occur. Common causes of bilateral

---

Vital Signs and Resuscitation, by Joseph V. Stewart. ©2003 Landes Bioscience.

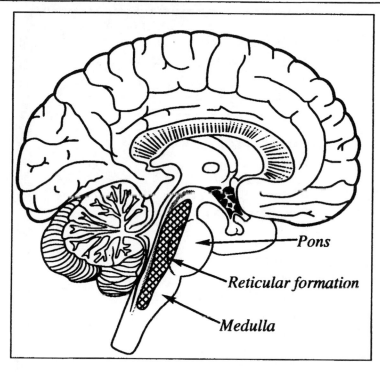

Pons

Reticular formation

Medulla

Fig. 6.1. The Reticular Formation.

cortical involvement are drugs, and oxygen or glucose deficits (**metabolic etiology**).

The brainstem consists of the thalamus, hypothalamus, midbrain, pons and medulla, and contains centers for vital functions. The **reticular formation**, an inner core of neurons in the spinal cord and brainstem, regulates respiration, blood presssure, heart-rate, endocrine secretion, conditioned reflexes, learning and consciousness. Incoming stimuli are integrated by the reticular formation. A portion of it, the **reticular activating system (RAS)**, is responsible for the arousal reaction. Most patients with lesions of the RAS are comatose. Common causes of brainstem involvement are trauma and stroke (**structural etiology**). Some drugs affect the brainstem. For example, epinephrine, amphetamines and cocaine stimulate RAS conduction; opiates and barbiturates depress it.

In summary, a decrease in level of consciousness occurs if:

1. Both cerebral hemispheres are involved, or
2. The RAS brainstem is affected.

EYE OPENING:
    spontaneous 4
    to voice 3
    to pain 2
    none 1

VERBAL RESPONSE:
    oriented 5
    confused 4
    inappropriate words 3
    incomprehensible sounds 2
    none 1

MOTOR RESPONSE
    obeys commands 6
    purposeful movements (pain) 5
    withdraw (pain) 4
    flexion (pain) 3
    extension (pain) 2
    none 1

Score: 3 to 15
Comatose: 8 or less

Fig. 6.2. Glasgow Coma Scale (GCS).

The vast majority of patients with decreased level of consciousness (about 85%) are in the metabolic category.

Glasgow Coma Scale (GCS)
Teasdale and Jennett developed their scale in 1974 not only to assess changes in levels of consciousness in brain-damaged patients in prolonged comas, but also to avoid imprecise terms such as lethargy, semicomatose, stupor, etc. when describing the mental status of a patient (see Chapter 1). An important point about the GCS is that it can be used as a continuum for changing levels of consciousness. At about a GCS score of 8 the definition of coma is fulfilled. The GCS is an integral part of **trauma scales** (see Fig. 6.3).

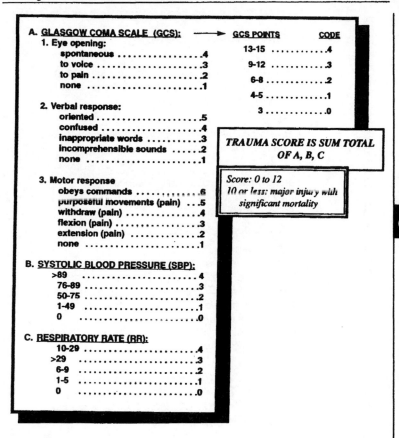

Fig. 6.3. Revised Trauma Score (RTS).

The GCS is sometimes bypassed in the field for AVPU, an easier method of evaluating level of consciousness because no numbering system is required.

A—alert

V—responds to verbal stimuli

P—responds to painful stimuli

U—unresponsive

The GCS is preferred over AVPU since both take about the same amount of time and a tenuous area exists between V and P in terms of airway protection. Sometimes it is difficult to assess a gag reflex. **Airway protection (endotracheal intubation) to avoid aspiration of vomitus is required when the gag reflex is lost or the GCS is 8** (Fig. 6.2).

Primary Survey/Resuscitation:

Airway/C-spine:
open airway (jaw thrust, chin lift), oxygen, secure airway (intubation, cricothyrotomy), pulse oximetry, clear c-spine as soon as possible

Breathing:
bag-valve-mask, ventilator, needle decompression, chest tube placement, treat pulmonary edema

Circulation:
cardiac monitor, immediate IV access
pulse --- treat dysrhythmias
shock --- treat

Disability: neurological exam

Immediate Therapy:
chemstrip **glucose** (if unavailable, D-50 1 amp, peds: D-25 4 ml/kg), **thiamine** 100 mg, **naloxone** 2 mg IV (peds: 0.1 mg/kg) (other therapy depends on Primary Survey, historical information, labs, also Secondary Survey (i.e. therapy for increased intracranial pressure, activated charcoal, IV fluids, packed cells)

Labs:
CBC, Chem 7, EKG (cardiac enzymes), ABG, chest x-ray, UA (Foley cath), tox screen (and ETOH), LFT's, NH_4, CO, head CT, (other labs as indicated, incl. trauma x-rays)

Secondary Survey:

Complete physical exam
History

Fig. 6.4. Management of the Comatose Patient.

Management of Altered Level of Consciousness

In contrast to the traditional approach in medicine, the comatose patient or the patient with a significant alteration in level of consciousness requires immediate management before completing the physical exam and acquiring the history.

The ABCs of resuscitation are followed (Fig. 6.4). When an immobilized patient arrives in the emergency department, the cervical collar and backboard are left in place until a cause is found for the decrease in level of consciousness. **Naloxone (Narcan)** 2 mg and **thiamine** (vitamin B-1) 100 mg are administered intravenously. If a fingerstick blood sugar is low or unavailable, glucose (50 cc of 50% dextrose) is administered after thiamine to reverse hypoglycemia.

Naloxone reverses the effects of a narcotic by competitive inhibition at the opioid receptor site. Thiamine prevents **Wernicke's Encephalopathy**, a rare neurological condition caused by thiamine deficiency seen in alcoholics with poor nutrition. Signs and symptoms include nystagmus, occular nerve palsy, ataxia and confusion. Thiamine functions as a coenzyme in the breakdown of glucose. Glucose given before thiamine depletes what little thiamine is available for glucose metabolism and may precipitate the syndrome. Glucose and thiamine may be administered at the same time.

Naloxone, thiamine and glucose were referred to in the past as a "coma cocktail" and were often automatically administered. If a fingerstick glucose is normal, administering glucose is not indicated. The same applies to thiamine in the pediatric population. If a drug overdose is suspected, activated charcoal is administered by gastric tube after endotracheal intubation (Fig. 6.4).

Increased Intracranial Pressure

Increased intracranial pressure is a life-threatening event and must be dealt with immediately. Common causes are a head injury with intracranial bleeding and a hemorrhagic stroke. Signs include papilledema, loss of spontaneous venous pulsations, an increase in systolic pressure, bradycardia, an abnormal respiratory pattern and a fixed dilated pupil.

Carbon dioxide is a potent vasodilator in the brain and hyperventilation blows off carbon dioxide and reduces pressure. Mannitol is an osmotic diuretic that removes excess fluid from the brain. Increased intracranial pressure blocks blood flow to the brain, and the hypoxia triggers an increase in systolic pressure to re-establish flow. The increased blood pressure causes a baroreceptor decrease in heart rate, and pressure against the RAS of the pons and medulla decreases the heart and respiratory rates. The triad of increased blood pressure, decreased heart rate and irregular breathing is the **Cushing reflex**. In adults, often only the blood pressure rises. The triad occurs more often in pediatrics. Intracranial pressure may cause the brain to push against the third cranial nerve on that side causing a fixed dilated pupil, indicating compression of the lower part of the temporal lobe (uncus) against the tentorium cerebelli with impending herniation (Fig. 6.5).

Treatment: intubation, hyperventilation, the head of the bed is raised 30° (except in the trauma patient with a cervical collar), furosemide 40 mg

Common causes of sudden increase in ICP:
  head trauma with intracranial bleeding
  hemorrhagic stroke

Signs of increased ICP:
  headache
  decreased level of consciousness
  loss of spontaneous venous pulsations
  papilledema
  increased systolic pressure
  bradycardia
  irregular breathing
  fixed dilated pupil

Treatment for increased ICP:
  elevate head of bed to 30 degrees (first clear c-spine)
  intubation (rapid sequence may be required)
  hyperventilation to CO_2 of 30 mmHg
  mannitol 1 gm/kg IV over 5 min (consult neurosurgeon)
  furosemide 0.5 mg/kg IV (consult neurosurgeon)

Fig. 6.5. Increased Intracranial Pressure (ICP).

IV and/or mannitol 1 gm/kg IV is administered in consultation with a neurosurgeon.

Neurological Examination

Signs of Metabolic Injury

Signs of metabolic injury, implying an intact brainstem, are roving eye movements, a pupillary reaction to light (pinpoint pupils suggest opiates or a pontine lesion. Dilated reactive pupils are seen with adrenergic or anticholinergic drugs), a normal oculocephalic reflex (doll's eyes) consisting of abruptly rotating the head to one side while the eyes deviate in the opposite direction (this test should not be used in the trauma patient unless the c-spine has been cleared), a normal **oculovestibular reflex** (instilling 50 ml of cold water into the auditory canal causes deviation of the eyes toward the water) and hyporeflexia.

Signs of Structural Injury

Signs of structural injury from trauma or stroke are fixed pupils, either large or pinpoint (pinpoint pupils suggest a pontine hemorrhage. Fixed mid-position pupils implies brainstem damage. One fixed dilated pupil suggests impending uncal herniation), no extra-ocular movements, **loss of** oculocephalic and **oculovestibular reflexes**, differences in movements of arms and legs, asymmetry and increased deep tendon reflexes with upgoing toes (**Babinski's**

SIGNS SUGGESTING METABOLIC CAUSE:
1. **Pupils that react to light (some exceptions)**
2. **Roving eye movements**
3. **Normal oculocephalic/oculovestibular reflexes (brainstem intact)**
4. **Hyporeflexia**

SIGNS SUGGESTING STRUCTURAL CAUSE:
1. **Signs of trauma**
2. **Fixed pupils**
3. **Absent oculocephalic/oculovestibular reflexes (brainstem injury)**
4. **Hyperreflexia/focal signs/up-going toes**
5. **Decerebrate posturing/decorticate posturing (rarely metabolic)**

Fig. 6.6. Metabolic vs. Structural Signs of Coma.

reflex) and **decorticate** or **decerebrate posturing** (arm flexion and leg extension in decorticate posturing represents injury to both cerebral hemispheres; extension of the arms and legs in decerebrate posturing represents injury to the brainstem). Decerebrate and decorticate posturing may occur in metabolic derangements, but more commonly are seen with structural damage. Fixed ocular deviation is toward a cortical lesion (Figs 6.7, 6.8).

Physical Examination

Vital signs may give a clue to the etiology. Hypothermia (including metabolic causes such as hypothyroidism, hypoadrenalism, hypoglycemia and sepsis) and hyperthermia may all cause a decreased level of consciousness.

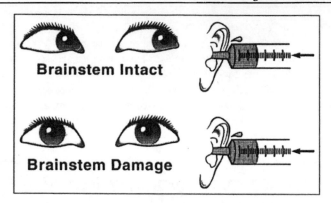

Fig. 6.7. Oculovestibular Reflex.

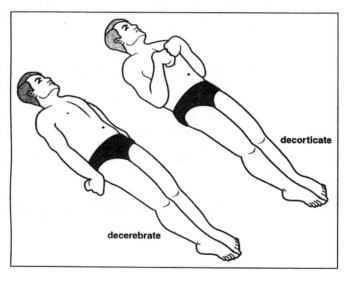

Fig. 6.8. Decerebrate vs. Decorticate Posturing.

Tachyarrhythmias and bradyarrhythmias suggests cardiac disease. Hyperventilation is seen in diabetic ketoacidosis, uremia and cirrhosis. Hypoventilation is common in opiate drug overdoses and in pulmonary disease. Hypotension requires searching for the etiology of shock (see Chapter 5). Hypertension suggests hypertensive encephalopathy or drugs such as cocaine (Chapter 5).

Breath: alcohol on the breath is noted, as is the fruity odor of diabetic ketoacidosis. A petroleum or garlic odor is sometimes seen in organophosphate pesticide poisoning.

HEENT: evidence of trauma should be sought, such as bruising or lacerations of the head and face. A basilar skull fracture may cause cerebrospinal fluid to leak from the nose or ear, or the extravasation of blood in the middle ear (hemotympanum) into the skin around the eyes (raccoon eyes) or over the mastoid process (Battle's sign).

Neck: neck stiffness, Kernig and Brudzinki's signs may indicate meningitis (see next section). Jugular venous distention is noted, as is the size of the thyroid gland.

Chest: signs of trauma should be sought, as above. The heart and lungs are evaluated. Decreased breath sounds on one side may indicate a pneumothorax or hemothorax. Jugular venous distention plus low blood pressure may suggest cardiac tamponade (see Chapter 5).

Abdomen: jaundice and a distended abdomen is seen in alcoholic liver disease. Abrasions, decreased bowel sounds, tenderness and rigidity suggest trauma with possible internal hemorrhage.

Neurologic: (see previous section) the Glasgow Coma Scale is assessed frequently for changes.

Skin: abrasions suggest trauma, jaundice suggests sequelae of liver disease, needle-tracks suggest drug abuse. Profuse sweating is seen with organophosphate pesticide poisoning. Cold sweats are present in a patient suffering a heart attack. A petechial or purpuric rash should alert one to meningococcemia.

Causes and Treatments of Coma

Since one cannot question the patient, other avenues for history are utilized. Medical tags and bracelets are sought. The questioning of EMTs is vital. Were empty medicine bottles present in the house? If so, they should be brought to the emergency department. Family and friends and neighbors and bystanders should be questioned. A handy mnemonic device for remembering the multiple causes of coma is "TIPS" and "AEIOU". **Treatment** for coma is supportive until the cause is found. The ABC's of resuscitation are strictly followed (Fig. 6.9).

"TIPS":

Trauma, Temperature

Trauma: in addition to head trauma, shock from hemorrhage, pericardial tamponade, myocardial contusion and tension pneumothorax may cause a decreased level of consciousness (see Chapter 5). A **concussion** is a transient loss of consciousness with no brain damage. A **contusion**, or bruising of the brain with small hemorrhages and tissue tears, usually causes a loss of

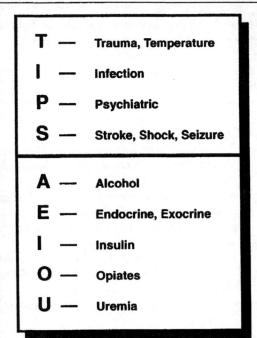

Fig. 6.9. Common Causes of Coma.

T — **Trauma, Temperature**

I — **Infection**

P — **Psychiatric**

S — **Stroke, Shock, Seizure**

A — **Alcohol**

E — **Endocrine, Exocrine**

I — **Insulin**

O — **Opiates**

U — **Uremia**

consciousness, sometimes briefly, sometimes for a long period (**diffuse axonal injury**). A **traumatic subarachnoid** hemorrhage from injury to vessels in the pia causes bleeding into cerebrospinal fluid in the subarachnoid space, sometimes producing headache and stiff neck. An **epidural hematoma** is a collection of blood between bone and dura from a laceration of the middle meningeal artery. A **subdural hematoma** is blood between the dura and arachnoid from tears in bridging dural veins. An intracerebral hemorrhage is the accumulation of blood within brain substance. A CT will not show concussions or in many cases contusions, but does reveal epidural, subdural and intracerebral hemorrhages. **Treatment:** increased intracranial pressure (ICP) is treated as previously described. Epidural and subdural hematomas require surgical evacuation (Fig. 6.10).

Temperature: hypo- and hyperthermia are discussed in Chapter 2.

Infection

Common infections causing decreased levels of consciousness are **sepsis** and **bacterial meningitis** (viral meningitis usually does not cause coma, except in the pediatric population). Neurological findings in sepsis range from lethargy or agitation to coma. Inflammatory mediators cause multi-

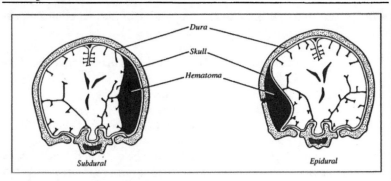

Fig. 6.10. Subdural and Epidural Hematomas.

organ system failure and hypotension (septic shock) with inadequate perfusion to the brain. Sepsis and septic shock are discussed in Chapter 5.

Bacterial meningitis is seen primarily in pediatrics and the elderly, with sporadic outbreaks in other populations. With the advent of the *H. influenzae* vaccine, the main organism is *Strep. pneumoniae*, not only in peds but in all age groups. Seeding is from bacteremia, otitis media and sinusitis. Fever, headache, altered mental status and HIV+ are important historical items. Seizures may occur. The physical exam in infants may show hypothermia, a bulging fontanelle, lethargy, dehydration and otitis media (see Chapter 7). Older children and adults usually have nuchal rigidity, pain on extension of the legs (**Kernig sign**) and passive neck flexion producing flexion of the hips (**Brudzinski sign**). In meningococcal meningitis the skin may show petechiae and purpura. **Treatment:** when meningitis is suspected, IV antibiotic therapy (ceftriaxone or cefotaxime 2 gm, 50mg/kg in peds) is begun before lumbar puncture. The presence of papilledema and loss of spontaneous venous pulsations indicate increased intracranial pressure, and therapy for ICP should be begun immediately (see earlier section). When ICP is suspected, a CT should be done before an LP.

Stroke, Shock, Seizures

A patient with a **cortical ischemic stroke** involving one side of the brain, with a profound sensory and motor loss on the opposite side of the body along with aphasia (left brain) or inattention and unconcern (right brain) and sometimes confusion, experiences no loss of consciousness unless massive ischemia causes brain edema. The less common **brainstem ischemic stroke** (basal artery) causes coma from involvement of the RAS (see earlier section). A **hemorrhagic stroke** begins with a headache and alteration in consciousness that progresses to coma because of a severe global mass effect with increased intracranial pressure, compression of the brainstem and herniation. Diagnosis is made by CT. **Treatment:**

1. The ABCs are followed (fig. 6.4);
2. Blood pressure over 220/120 is treated with increments of labetalol 20 mg IV;
3. Increased intracranial pressure is controlled (Fig. 6.5);
4. Neurosurgical consultation is obtained and
5. The thrombolytic t-PA 0.9mg/kg IV (maximum 90mg) may be given over an hour for an ischemic stroke if the time of onset is known to be less than three hours (and no contraindications exist).

A **subarachnoid hemorrhage** is caused by rupture of a congenital (berry) aneurysm in the Circle of Willis at the base of the brain, either at rest or during exercise. The patient describes a sudden **severe headache**. Bleeding into the subarachnoid space and ventricles produces a mild to severe decrease in level of consciousness. Preliminary diagnosis by CT or lumbar puncture may be supplemented by angiography. **Treatment:**

1. Blood pressure is controlled by labetalol 20mg IV increments in pre-hemorrhage levels;
2. The calcium-channel blocking agent nimodipine 60mg PO every 6 hours reduces vasospasm;
3. Seizures are prevented with fosphenytoin (cerebyx) 15mg/kg IV as a loading dose and
4. Neurosurgical consultation is obtained.

Shock is discussed in Chapter 5.

Seizures: the most **common cause of a seizure** is failure to take anticonvulsive medicine. Decreased level of consciousness is transient and the person gradually awakens (post-ictal state). A rapid blood sugar is checked and glucose is administered as needed. if the sezure continues, it is stopped with lorazepam (Ativan) 4 mg IV over 2 minutes (Peds: 0.1 mg/kg) (or midazolam ((Versed)) 0.2 mg/kg IM). For the persistent seizure (**status epilepticus**), a second dose of fosphenytoin (Cerebyx) 20 mg/kg IV. If no response occurs, phenobarbitol 18 mg/kg IV is used, and intubation may be required. The continuous seizure may require a neuromuscular blocking agent (i.e., vecuroium 0.1 mg/kg) or general anesthesia.

Other causes of seizures are congenital/genetic disorders, brain tumors, eclampsia (discussed in Chapter 5), drugs such as theophylline, phenothiazines, lithium, cocaine and antidepressants, opportunistic cerebral infections in AIDS patients and febrile seizures (discussed in Chapter 7).

"AEIOU":

Alcohol/Drugs

Alcohol: wide variability exists in each person's response to alcohol, depending on whether one is a chronic alcoholic or an occasional drinker. This results in various degrees of intoxication, physical **dependency** (withdrawal symptoms on stopping the drug) and **tolerance** (increased amounts of drug for the same effect). In the emergency setting, it is not uncommon to see an alert and oriented alcoholic with a blood alcohol level of 400 mg/dL, while a nonalcoholic may be comatose at that level. A level of 100 mg/dL is legal intoxication in most states. The nontolerant person usually shows a decrease in level of consciousness at a level of about 300 mg/dL. Coma (GCS of 8) may occur at about 400 mg/dL (often requiring intubation), and death from respiratory depression may occur at 500 mg/dL (LD-50).

The alcoholic is at increased risk for a subdural hematoma, and a search for bruises and abrasions should be sought. A rectal temperature is required. A low threshold should exist for a head CT, as well as a diagnostic peritoneal lavage to rule out abdominal injuries. Labs should include, in addition to a serum ethanol level, a drug screen (cocaine is a common accompanying drug) and a serum ammonia to rule out hepatic encephalopathy. Alcohol withdrawal is seen about 48 hours after the last drink and exhibits a wide variety of manifestations, including anxiety, tremors, visual hallucinations and seizures, but usually does not show a decreased level of consciousness. Withdrawal may be seen in the intoxicated patient. **Treatment** for alcohol withdrawal: one liter of D5NS with $MgSO_4$ 2 gm, folate 1 mg and an ampule of multivitamins is administered for both intoxication and withdrawal since glycogen, magnesium and vitamins are usually depleted. Gastric decontamination with lavage and charcoal is indicated only in the rare case of an acute ingestion of a large amount of alcohol over a short period of time in a nonalcoholic. For withdrawal, lorazepam (Ativan) 2-4 mg IV is administered, followed by 2 mg every 30 minutes as needed.

The alcoholic may present with liver failure and coma from **hepatic encephalopathy**, a condition in which nitrogenous and other compounds (i.e., ammonia, gamma-aminobutyric acid, mercaptans) normally removed by the liver accumulate and gain access to the central nervous system, causing neuroinhibition and cerebral edema. The serum ammonia is elevated, electrolytes are often abnormal and asterixis ("liver flap" = hand tremor) is sometimes present. **Treatment** for hepatic encephalopathy: fluid and electrolyte abnormalities are corrected. Lactulose may be given via nasogastric tube (30 cc TID). Lactulose is a nonabsorbable disaccharide when in contact with colonic bacteria traps ammonia in the colon as nondiffusible ammonium ions. Neomycin (1 gm via NG q8h) suppresses bacteria responsible for the production of ammonia and other nitrogenous compounds.

Drugs: it is not appropriate to list the multitude of drugs and toxins that may cause an altered level of consciousness. Common drugs causing coma are sedative/hypnotics and narcotics. An altered level of consciousness is seen occasionally with carbon monoxide poisoning. Narcotics (i.e., morphine, heroin) bind to opioid receptors in the brain (*mu* receptors). Sedative-hypnotics (alcohol, benzodiazepines and barbiturates) facilitate gamma-aminobutyric acid (GABA), the major inhibitory neurotransmitter of the central nervous system. Benzodiazepines and barbiturates bind to GABA receptors. No receptor has been identified for alcohol, but alcohol enhances the action of GABA receptors. **Treatment:** the treatment for an overdose of a sedative/hypnotic or narcotic is gastric lavage within the first hour. After the first hour **activated charcoal** 50 gm PO or by gastric tube is administered to prevent absorption of the drug. If the gag reflex is lost, the patient is intubated and gastric lavage and/or charcoal is administered by gastric tube. The antidote for opiate ingestion is **naloxone**, a competitive antagonist at the *mu* opioid receptor site. The person awakens quickly. Naloxone may then be administered 1 mg in 250 cc NS at 250 cc per hour. **Flumazenil** (Romazicon) 0.2 mg IV is a competitive antagonist of the benzodiazepine receptor and will reverse a benzodiazepine overdose. However, since seizures have been reported with Flumazenil, it should be used for diagnostic rather than therapeutic purposes.

A common inhalation injury is **carbon monoxide** poisoning from fires, faulty gas heaters and in machine-shops where ventilation is poor. Carbon monoxide (CO) is odorless and binds to hemoglobin 210 times more readily than oxygen. Early symptoms are headache, dizziness, weakness and nausea. Consciousness is affected at CO levels of 20-30%, confusion and syncope occur at 40%, coma and seizures take place at 50%, and death occurs at 60%. **Treatment:** 100% oxygen by nonrebreather mask until the level is 0. Use of hyperbaric oxygen therapy (HBO) is controversial, but is indicated for comatose patients, those with major illnesses such as coronary artery disease and the pregnant patient.

Endocrine, Electrolytes

Endocrine: severe **hypothyroidism** (myxedema coma) and acute adrenal insufficiency (from an exacerbation of Addison's disease) are occasional endocrine causes of decreased level of consciousness. Signs of myxedema are hypothermia (75%), obesity, a surgical neck scar, cool dry skin, delayed DTRs, anemia, electrolyte disturbances, and occasionally respiratory distress, bradycardia and CHF. **Treatment:** hydrocortisone 100 mg and levothyroxine 4 µg/kg IV. Acute **adrenal insufficiency** is sometimes seen when steroids are withdrawn in a patient with Addison's disease. Patients present with confusion, lethargy, nausea, vomiting, hyperpigmentation and hypotension. Lab studies show hyperkalemia, hyponatremia and hypoglycemia. **Treatment:**

hydrocortisone 100 mg and 2 liters of D5NS to restore blood pressure and correct hypoglycemia. Hyperkalemia is treated with an ampule of bicarbonate over 5 minutes. If the K+ level is >8 meq, 10 cc of a 10% calcium gluconate solution is administered over 10 minutes.

Electrolyte imbalances rarely cause a decrease in level of consciousness. Occasionally severe hyponatremia may cause decreased mentation from the syndrome of inappropriate antidiuretic hormone (SIADH) secretion from a previous intracranial lesion or tumor. **Treatment:** water restriction. For severe hyponatremia (< 115 mEq), 3% saline is administered at 100 cc per h.

Insulin

The most common disorder causing an altered and/or decreased level of consciousness in all categories is **hypoglycemia** (glucose level < 50 mg) in the insulin-dependent diabetic patient with decreased food intake (insulin reaction) or the occasional patient on oral medication. Confusion is present, and if sustained for any length of time a decreased level of consciousness occurs. **Treatment:** IV glucose (D50) quickly reverses the condition. If an IV is difficult to obtain, glucagon 1-2 mg may be administered IM (glucagon catabolizes stored glycogen). **Diabetic ketoacidosis (DKA)**, indicated by glucose >250 mg, bicarb <15 mEq and pH <7.3, is treated with normal saline, 5 L at 500 cc per hour, insulin 0.2 units/kg initially, then 0.1 unit/kg/h IV infusion, potassium 20 mEq per liter and bicarbonate 1 to 2 amps if pH is <6.9. The patient with **hyperosmolar hyperglycemic nonketotic coma** (HHNC) (glucose >600 mg, negative ketoacidosis) is lethargic and confused but rarely comatose. Fluid and potassium treatment is similar to DKA, except that less insulin and more fluid is required.

Oxygen

Oxygen deficits are discussed in Chapter 4. An insidious cause of decreased mentation is **hypercapnia** (carbon-dioxide narcosis) from hypoventilation from many causes: two common ones are acute pulmonary edema and a tiring asthmatic. The pCO_2 level in these cases is elevated above 50 mmHg (sometimes to 100 mmHg) and the pH is decreased (respiratory acidosis). Pulse oximetry may show a normal oxygen saturation. **Treatment:** increased ventilatory support, requiring intubation and mechanical ventilation.

Uremia

Uremia (chronic renal failure) is caused by the accumulation of toxic products of protein metabolism in the bloodstream. The blood urea nitrogen (BUN) and creatinine are increased. An altered mental status is believed to be related to loss of the kidney's capacity to synthesize vitamin D, leading to hypocalcemia and secondary hyperparathyroidism. Definitive treatment is dialysis.

Practical Points

- First, the ABCs of resuscitation are followed (see Chapter 8).
- Everyone who deals with critically ill patients should be familiar with the Glasgow Coma Scale. With the exception of AVPU, it is the simplest of all the scores to remember, and thus it is the most useful. A GCS of 8 is the definition of coma.
- In the critical patient, and particularly the trauma patient, serial vitals and GCSs are important.
  Examples:
  1. LOC: rollover MVA with prolonged extrication. No eye opening to pain, incomprehensible sounds, no response to painful stimuli: GCS 4. En route 5 minutes later—eye opening to voice, confused, withdraws from pain: GCS 11.
  2. LOC: possible overdose. Eyes open to verbal, confused, obeys commands: GCS 13.

References

1. American College of Surgeons. Advanced trauma life support (ATLS). Chicago, 1997.
2. Braakman R et al. Prognosis and prediction of outcome in comatose head injured patients. Acta Neurochir Suppl 1986; 36:112.
3. Charness M et al. Ethanol and the nervous system. N Engl J Med 1989; 321:7.
4. Chuidian F. The unconscious patient: Evaluation and first-line interventions. J Crit Ill 2000; 15:14.
5. Diamond I. Alcohol neurotoxicity. In: Asbury A et al. Clinical Neurobiology. Philadelphia:WB Saunders, 1992.
6. Hemphill R et al. Delayed presentation after head injury: Is a computed tomography necessary? Acad Emerg Med 1999; 6:957.
7. Huff J. Altered mental status and coma. In: Tintinalli J et al. Emergency Medicine: A Comprehensive Study Guide. New York: McGraw-Hill, 2000.
8. Jackimczyk K. Altered mental status and coma. In: Markovchick V ed. Emergency Medicine Secrets, Philadelphia: Mosby, 1999.
9. Jennett B. Clinical assessment of consciousness. Acta Neurochir Suppl 1986; 36:90.
10. Negrini B et al. Cerebrospinal fluid findings in aseptic versus bacterial meningitis. Pediatrics 2000; 105:316.
11. Plant J, MacLeod D. Response of a promethazine-induced coma to Flumazenil. Ann Emerg Med 1994; 24:5.
12. Plum F et al. Disorders of consciousness and higher brain function. In: Andreoli T et al. Cecil Essentials of Medicine. Philadelphia: WB Saunders, 1997.
13. Scheinkstel C et al. Hyperbaric or normobaric oxygen for acute carbon monoxide poisoning. Med J Aust 1999; 170:203.
14. Starkman S. Altered mental status. In: Hamilton G, ed. Emergency Medicine—An Approach to Clinical Problem-Solving. Philadelphia: WB Saunders, 1991.
15. Starmark J, Lindgren S. Is it possible to define a general "conscious level"? Acta Neurochir Suppl 1986; 36:103.
16. Sternbach G. The Glasgow Coma Scale. J Emerg Med 2000; 19:67.
17. Teasdale G, Jennett B. Assessment of coma and impaired consciousness: A practical scale. Lancet 1974; 2:81.

CHAPTER 7

Pediatric Vitals

The APGAR Score

The first vitals are recorded at birth. In 1952, Virginia Apgar, an anesthesiologist at the Sloane Hospital for Women in New York City, after assisting in the delivery of >17,000 babies, developed a scoring system to assess asphyxia and predict neurologic outcomes in newborns. The scale, named after her, consists of five parameters: heart-rate, respirations, reflex irritability, muscle tone and color. A score is assigned at one minute and five minutes after birth. Most newborns have a score between 7 and 10; from 4-6 is moderately depressed, and 0-3 is severely depressed requiring resuscitative measures. If the 5 minute score is less than 7, scores are done every 5 minutes for 20 minutes. Apgar's last name has been used as an acronym for remembering the system:

A—Appearance (color)
P—Pulse
G—Grimace (reflex irritability)
A—Activity (muscle tone)
R—Respirations

Resuscitative measures for low APGAR scores is discussed in Chapter 8 (Fig. 7.1).

Sign	0	1	2
Heart rate	Absent	Slow (< 100/min)	>100/min
Respirations	Absent	Slow, irregular	Good, crying
Muscle tone	Limp	Some flexion	Active motion
Reflex irritability (catheter in nares)	No response	Grimace	Cough or sneeze
Color	Blue or pale	Pink body with blue extremities	Completely pink

Fig. 7.1. APGAR Score.

Vital Signs and Resuscitation, by Joseph V. Stewart. ©2003 Landes Bioscience.

Temperature

The Newborn

The temperature of the newborn is normally the same as that of the mother. Infants, particularly newborns, are prone to hypothermia. The high ratio of body surface area to body mass causes 4 times more heat loss by radiation and evaporation than in the adult. A contributing factor is the sparse insulation against heat loss from the developing keratinization of skin and subcutaneous fascia. Babies do not shiver. Instead they respond by secreting catecholamines which constrict vessels and mobilize brown fat. The mobilization of brown fat, which contains mitochondria that hydrolyze and oxidize free fatty acids for energy, increases the metabolic rate by two-fold or more. Impaired tissue perfusion from the cold may result in metabolic acidosis, shock and cardiac arrest. The incubator or overhead heater maintains the infant at a temperature of 97.7-98.6 F°(36.5-37°C).

Pediatric Temperature

Pediatric temperatures fluctuate, but generally parallel adult readings. For a quick-screening in the newborn and infant, a heat sensitive strip containing liquid crystals that change color as the temperature changes may be applied to the forehead and a readout recorded. However, these are often inaccurate. A recent temporal artery thermometer which measures forehead temperatures showed inconsistent results in the birth to 12-year-old group. In the neonate, because of the unique distribution of body fat, the axillary temperature is sometimes used. The thermometer should remain in place for at least 5 minutes. **Rectal temperatures** should be routinely done on infants and small children. The infant or small child is positioned on his back and the thighs and knees are flexed while he is held, or prone on the mother's lap with hips flexed. The probe is inserted 1 inch for children and 1/2 inch for infants. Use of the tympanic thermometer is controversial in the period up to 3 months. The probe must make a tight seal for accuracy (see Chapter 2). Unfortunately, since the auditory canal is small and current probe sizes fairly large, inappropriate caution by the user creates inconsistent and thus unreliable results.

Fever (see also Fever section, Chapter 2)

In the early pediatric age group, the potential exists with fever and a high white count for a severe infection to be present. During the first few months of life, the immune system is developing and the infant is protected from infection by antibodies from the mother (maternal antibodies—passive immunity). Passive immunity fails occasionally. Infants less than three months of age with core (rectal) temperatures of 101.3F/38.5C have twenty times more risk of serious infection than do older children (pediatric fever is defined as a rectal temperature of 100.4F/38C). If a source for the fever is found, the infant or child is treated appropriately. If no source is found (**fever without a source—FWS**), the workup proceeds for **serious bacterial infection (SBI)**, such as sepsis/meningitis.

In the **neonate (0-28 days)**, a temperature of 100.4F (38C) mandates hospitalization and a workup for SBI: CBC, blood culture, urine culture, lumbar puncture, chest x-ray. "Early onset" sepsis (0-5 days after birth) is usually caused by maternal transmission. Signs and symptoms may be entirely absent, or may include poor feeding, lethargy, respiratory problems, vomiting, tachycardia and hypothermia (more common than fever). **Treatment:** admission and antibiotic therapy such as IV cefotaxime (Claforan) 50mg/kg q 6h.

Infants in the **1 to 3 month age group** with FWS and high risk criteria such as a temperature of 100.4F (38C), a white blood count of 15,000 with bands or toxic signs such as seizure, a weak, shrill or continuous cry, a bulging fontanelle, lethargy, a hemorrhagic rash or paradoxical irritability (picking up and comforting a child usually stops crying—in paradoxical irritability, since the movement causes meningeal irritation and pain, picking up induces crying) require a workup for sepsis/meningitis (neonates may also exhibit some of the above characteristics). **Treatment:** A blood culture, urine culture and lumbar puncture are obtained and IV antibiotics are administered, such as ceftriaxone (Rocephin) 50 mg/kg IV q 12h. **"Low risk" Rochester and Philadelphia Criteria** are used today to manage nontoxic infants at home: perviously healthy term infant, nontoxic appearance, WBC between 5000-15,000/ml without bands, UA with <5 WBC/hpf and <5 WBC/hpf in stool if diarrhea present. Two outpatient options exist: 1) blood culture, urine culture, lumbar culture, ceftriaxone (Rocephin) 50 mg/kg IM and re-evaluation in 24 hours, or 2) blood culture, urine culture and re-evaluation in 24 hours.

Children in the **3 to 24 month age group** with FWS are less likely to have life-threatening illnesses than the 0 to 3 month group. A temperature of >102.2F (39C) requires a CBC, UA and chest x-ray. **Treatment:** Toxic children are admitted for septic workup and parenteral antibiotics. Nontoxic children with temperatures <102.2F (39C) may be sent home: if the WBC count is >15,000, a culture is done and ceftriaxone (Rocephin) 50mg/kg IM is administered. Outpatient antibiotics are prescribed as appropriate for otitis media (i.e., amoxicillin 40mg/kg/day TID or erythromycin/sulfisoxazole 50mg/kg/day TID for 10 days, pneumonia (i.e., amoxicillin 40mg/kg/day TID for 10 days or azithromycin 10mg/kg initially, then 5mg/kg/day for four days) or UTI (ie., trimethroprim-sulfamethoxazole 8mg/kg/day BID for seven days).

Febrile seizures occur in the five month to five year age group (most under age two). They usually do **not** occur at extremely high temperatures, but rather at core temperatures of about 102F (38.9C), are usually of benign etiology and often have a genetic component. In the average case, the seizure is generalized, lasts less than 5-10 minutes (often less than one minute) and behavior returns to normal in less than an hour. A source for the fever is sought (i.e., upper respiratory infection, otitis media, pneumonia, urinary tract infection). **Treatment:** if a source is found, the patient is treated for that condition. If none is found, the patient is treated for FWS described above. In the under-one-year age group, a lumbar puncture is sometimes performed for a first time seizure and treatment is based upon the result. In the older child, a CBC and blood culture are performed, the patient receives appropriate antibiotic therapy, and is sent home with close follow-up.

Heart Rate/Pulse

The newborn heart rate is about 140 beats per minute, and gradually decreases to adult values at about age 14. Abnormal pediatric rhythms are rare. Sinus tachycardia is often seen in dehydration and/or infection (i.e., sepsis—see later section).

Supraventricular tachycardia is the more common of the rare arrhythmias in pediatrics. As in the adult, it is usually caused by a re-entry mechanism. The rate in infants is usually greater than 220 beats per minute, and in children greater than 180. **Treatment:** In the stable patient, vagal maneuvers are attempted and adenosine (Adenocard) 0.1 mg/kg is administered as a bolus. If this is not effective, a second dose is administered at 0.2 mg/kg. In the unstable patient or if adenosine is ineffective, synchronized cardioversion is performed at 0.5 joules/kg (see Fig. 7.3).

Bradyarrhythmias in children are usually the result of hypoxemia and/or acidosis, and treatment is directed at correcting the underlying cause. Unstable patients may require epinephrine 0.01 mg/kg (1:10,000) every five minutes. Atropine 0.02 mg/kg is administered if the response to epinephrine is poor after two doses. Cardiac pacing may be required (Fig. 7.4).

Respiration

The respiratory rate is about 60 for the first two days after birth, then decreases to about 40. Adult values of 12 are reached by about age 14 (Fig. 7.2). In infants, periodic breathing is sometimes seen where respiratory pauses exist because of a lack of complete development of respiratory control.

Rarely, an infant may have an episode of prolonged apnea (>20 sec) sometimes accompanied by choking and gagging. This "**apparent life-threatening event**"—**ALTE** (also known as "near-miss SIDS") is a respiratory problem that appears to be related to **sudden infant death syndrome (SIDS)**. It is seen between the ages of 1 month and 1 year (with peaks at 2 and 4 months). Over 70 theories have been proposed for ALTE and SIDS. Among the more substantive are prematurity, sleeping prone, siblings with SIDS, substance-abusing mothers, respiratory syncytial virus, child abuse, gastroesophageal

AGE		AVERAGE PULSE		AVERAGE RESPIRATIONS		AVERAGE SYSTOLIC BP
NEWBORN	—	140	—	30-60	—	70/55
1-6 months	—	130	—	30-40	—	85/55
6-12 months	—	115	—	24-30	—	90/55
1-4 years	—	108	—	20-30	—	96/60
4-6 years	—	102	—	20-25	—	100/60
8-12 years	—	94	—	16-20	—	100/65
>12 years	—	84	—	12-16	—	110/70

Fig. 7.2. Pediatric Vital Signs.

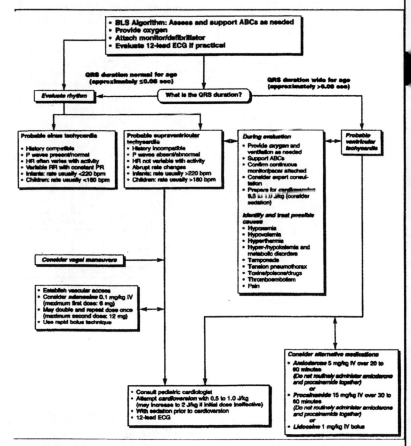

Fig. 7.3. Pediatric Tachycardia Algorithm. Reprinted with permission from: Guidelines for 2000 for Cardiopulmonary Resuscitation and Emergency Cardiovascular Care, American Heart Association.

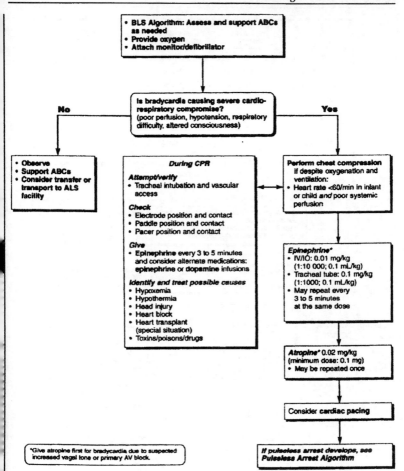

Fig. 7.4. Pediatric Bradycardia Algorithm. Reprinted with permission from: Guidelines for 2000 for Cardiopulmonary Resuscitation and Emergency Cardiovascular Care, American Heart Association.

reflux and infantile botulism. **Treatment:** hospitalization for an apnea workup. The infant is then sent home with an apnea monitor.

Upper Respiratory Emergencies

Obstruction (complete obstruction is discussed in Chapter 8)
Partial obstruction by a foreign body in the child may pose a problem because the circumstance is often unwitnessed and signs may be confusing. Choking, coughing and gagging may occur, then subside as the object passes into a smaller airway, usually the right mainstem bronchus (the anatomical

continuation of the trachea). This may later produce coughing, wheezing or stridor in any combination (a foreign body in the upper esophagus causes stridor, drooling and dysphagia). The diagnosis is made by a high index of suspicion and various x-ray techniques, among them bilateral decubitus chest x-rays. The normal chest shows decreased relative volume on the downside compared to the upside. With obstruction on the downside, the downside remains fully inflated. **Treatment:** laryngoscopy or bronchoscopy with removal of the object in the operating room under anesthesia. Esophageal foreign bodies are removed by endoscopic forceps (see also Chapter 8).

Croup

Croup, or laryngotracheobronchitis, is a viral infection (parainfluenza virus) of the upper airway in the 6-month to 3-year-old population (most under 1 year). Fever and a barking cough are present, mostly at night, often accompanied by mild stridor, tachypnea and retractions. **Treatment:** humidified oxygen ("mist wand"), racemic epinephrine 0.5 cc in 3 cc NS, prednisolone (Prelone syrup) 1 mg/kg PO or dexamethazone (Decadron) 0.3 mg/kg PO or IM and PO fluids.

Epiglottitis

Epiglottitis is an infection of supraglottic tissue causing edema and partial obstruction of the glottis (airway). The incidence has decreased because of *Hemophilus influenza* vaccine. Other organisms are Strep and Staph. The median age is 7. Symptoms are a sudden onset of fever, sore throat and difficulty swallowing and talking (dysphonia). Stridor and drooling are often present, and the child sits with the chin forward. **Treatment:** because obstruction of the upper airway may occur at any time, a portable neck xray is taken in the emergency department, intubation equipment is assembled, and the doctor remains with the patient at all times. If the x-ray confirms the diagnosis, arrangements are made for immediate intubation by anesthesiology in the operating room. The child remains seated, and no blood is drawn or other treatments begun. If the child is seen in a doctor's office, a physician able to intubate should accompany the child to the hospital. After the airway is secured, cultures are taken and antibiotics are administered (i.e., ceftriaxone 50 mg/kg every 24 hours).

Lower Respiratory Emergencies

Asthma

Asthma, discussed in Chapter 4, is an allergic reaction precipitated by irritants, stress and infection. Mast cells release histamine and other mediators which cause bronchoconstriction, followed by airway edema. Signs and symptoms include shortness of breath, cough, wheezing and tachypnea. Evaluation is by means of pulse oximetry, peak flow, and blood gases.

Treatment: oxygen is placed if the O_2 saturation is <94%, a beta-agonist such as albuterol 2.5 mg in 2 cc NS is administered by nebulizer over an hour, and prednisolone (Prelone syrup) 1 mg/kg is given orally.

The need for close observation is indicated with a history of frequent hospitalizations, a previous intubation, use of steroids, an oxygen saturation <94% and a peak flow less than 10% improvement after several nebulizer treatments. An oxygen saturation less than 90% mandates ABGs. A CO_2 of >50 mmHg indicates impending respiratory failure. Signs of respiratory distress are nasal flaring, retractions, agitation, lethargy, confusion, sweating, altered consciousness and grunting (see Pediatric Respiratory Failure, Chapter 8).

Bronchiolitis

Bronchiolitis is similar to asthma except that the cause is a virus (usually respiratory syncytial virus—RSV) causing an inflammation of the bronchioles. It is seen in a younger age group (under 2 years of age) than the asthmatic. Fever, wheezing and tachypnea are present. Evaluation is by oxygen saturation and ABGs. Nasal washings may identify RSV. **Treatment:** nebulized racemic epinephrine (as for croup) is administered. Steroids are not used. PO or IV hydration is helpful. Pulse oximetry showing an oxygen saturation of 91% or below and/or sustained tachypnea (respiratory rate of 60 or greater) indicates the need for admission.

Blood Pressure

Blood pressure is measured by auscultation, palpation, Doppler or an oscillometric instrument such as the Dinamap. The blood pressure cuff should be long enough to encircle the arm completely, with slight overlap. The width should be about 2/3 the length of the arm. The lower limit of normal blood pressure in a child may be estimated by the formula: 70 + (2 x the age in years). Example: a three year old should have a systolic pressure above 76 mm Hg (Fig. 7.5).

Shock

Hypovolemic Shock

Hypovolemic shock from dehydration caused by gastroenteritis is an important cause of shock in the pediatric age group. Hypovolemic shock may also be caused by blood loss from trauma. Dehydration is seen after continuous vomiting and/or diarrhea and by decreased fluid intake over several days. Signs and symptoms include tachycardia, tachypnea, decreased urinary output, altered mental status and dry mucous membranes (see Fig. 7.6). It is important to note that only in the final stages of shock does the blood pressure fall. Tachycardia is prevalent throughout all stages. The child is positively orthostatic if an increase in heart rate (>25 beats per minute

CUFF	WIDTH	LENGTH
NEWBORN	2.5-4.0 cm	5.0-10.0 cm
INFANT	6.0-8.0 cm	12.0-13.5 cm
CHILD	9.0-10.0 cm	17.0-22.5 cm

Fig. 7.5. Pediatric Blood-Pressure Cuffs.

Sign	Mild (5%)	Moderate (10%)	Severe (15%)
Pulse	Increased	Increased	Increased (weak)
Blood pressure	Normal	Normal	Decreased
Capillary refill	<2 sec.	>2 Sec.	>2 Sec.
Mucous membranes	Normal	Dry	Very Dry
Tears	Present	Decreased	Absent
Skin turgor	Normal	Decreased	Tenting
Urine output	Normal	Decreased	Absent
Fontanelle	Normal	Normal to sunken	Sunken

Fig. 7.6. Pediatric Dehydration.

rather than 30 as in the adult) or near-syncope occurs (see Chapter 5 Orthostatic Vital Signs, and Fig. 5.6). A good early indicator of shock in infants is **capillary refill**, discussed in Chapter 5. The fingernail bed is gently pressed and the time noted for the blanched nailbed to return to normal. Normal is less than 2 seconds. 2-3 seconds represents 5-10% dehydration. Longer than 3 seconds represents a greater than 10% deficit. Treatment for dehydration:

1. The ABCs of resuscitation are followed;
2. Normal saline 20 cc/kg IV bolus x 2 is administered; then
3. D5.25NS IV as maintenance, plus extra fluid to compensate for hypovolemia (Fig. 7.7) and
4. Fluids are adjusted so that the urine output is maintained at 1 ml/kg/hour.

Hypovolemic shock from blood loss: signs and symptoms are similar to those in dehydration. **Treatment:** the ABCs of resuscitation are followed, NS 20 cc/kg IV bolus x 2 is administered, the urine output is maintained at 1 cc/kg per hour, 10 cc/kg of type-specific warmed packed red blood cells

A. Initial Fluid: 2 boluses of normal saline
    at 20cc/kg.

B. Maintenance Fluid According to Weight:
    1. 0 - 10 kg = 100cc/kg per day
       (or 4cc/kg/hour)
    2. 10 - 20 kg = 1 liter for first 10 kg
       + 50 cc for each kg over 10kg/day
       (or 2cc/kg/hour)
    3. 20 + kg = 1½ liters for first 20 kg
       + 20 cc for each kg over 20kg/day
       (or 1cc/kg/hour)

C. Dehydration:
    5% = add 50cc/kg/day
    10% = add 100cc/kg/day
    15% = add 150cc/kg/day

**(NOTE: after initial boluses of normal
    saline, D5.25NS is used)**

Fig. 7.7. Pediatric Fluid Resuscitation.

(or O-negative RBCs if the need is urgent) is given, and a surgical consult is obtained as soon as possible (Figs. 7.6, 7.7).

Septic Shock

Septic shock is the last stage of a continuum from sepsis (see Fever, this Chapter, and Septic Shock, Chapter 5). Signs and symptoms include irritability, poor feeding and lethargy. Fever is present in the early stage, accompanied by tachycardia, tachypnea and warm and pink extremities. In later stages, inflammatory mediators are activated, pulses are weak, extremities are cool, mental status is decreased, capillary refill is prolonged, hypothermia may be present and the pulse pressure widens (hypotension is not seen until late, unlike the adult). The WBC may be high or low. Cultures are done on blood, urine and cerebrospinal fluid. **Treatment:**
    1. The ABCs of resuscitation are followed;
    2. 20 cc/kg boluses of IV NS are administered to maintain a urine
       output of 1 cc/kg per hour;

3. If BP can not be maintained, dopamine 5 ug/kg/min is added and
4. For the neonate, cefotaxime (Claforan) is given 50 mg/kg IV every 6 hours, plus ampicillin 50 mg/kg every 6 hours. For the infant and child, ceftriaxone (Rocephin) is administered at 50 mg/kg IV every 12 hours. Antibiotics should be administered when blood cultures are drawn, before the lumbar puncture.

Anaphylactic shock is discussed in Chapter 5.

Level of Consciousness

Basic features of decreased level of consciousness are discussed in Chapter 6. In contrast to the adult, the more common causes of decreased mental status in pediatrics are meningitis, poisoning and head trauma (common causes in adults are alcohol and drug abuse, hypoglycemia and hemorrhagic stroke). Drug ingestion has emerged as an important factor in the pediatric population. Although the Glasgow Coma Scale is useful for an older pediatric population, the "verbal response" section of the scale was restructured for children under 4 years of age. Recently a further modification was made for infants ages 0-23 months. Score totals are the same as for the adult. The **Pediatric Glasgow Coma Scale** has also been incorporated into the **Revised Trauma Score**. Steps in management are the ABC's, the administration of naloxone and glucose (if there is difficulty obtaining a blood sugar) and a low threshold should be present for lumbar puncture and head CT scanning (see Fig. 6.4). Increased intracranial pressure is treated without delay (Fig. 6.5). As with the adult, lab tests may be helpful, including an ammonia and toxicological studies. An important difference from the adult is that the history is usually easier to obtain (from parents—medical history, possible drug ingestion, trauma). The mnemonics TIPS and AEIOU are still useful (Fig. 6.9).

Common Causes of Pediatric Coma: TIPS, AEIOU (see also Chapter 6)

Trauma, Temperature: see also Chapters 2 and 8

Infection

An important cause of coma is **meningitis** from bacteremic invasion of the brain, causing inflammation and increased intracranial pressure. Two-thirds of all cases of meningitis are in pediatrics, most cases occurring between birth and age 2. *Strep pneumoniae* and *N. meningitidis* are the usual pathogens because of the recent use of *Hemophylus influenzae* type b (Hib) vaccine. A common presentation is fever accompanied by an altered mental state. Infants may show poor feeding, vomiting, paradoxical irritability (explained in the Fever section) and have a bulging fontanelle. Seizures are

EYE OPENING:
   Spontaneous 4
   To voice 3
   To pain 2
   None 1

BEST MOTOR RESPONSE:
   Spontaneous 5
   Localizes pain 4
   Withdraws to pain 3
   Flexion to pain (decorticate) 2
   Extention to pain (decerebrate) 1

Use only 1 of the following 2 categories:
BEST VERBAL RESPONSE
   (Age 2-5 yr)
         Appropriate words 5
         Inappropriate words 4
         Cries and/or screams 3
         Grunts .2
         No response 1

OR

BEST VERBAL RESPONSE
   (Age 0-23 mo)
      Smiles, coos, cries appropriately . .5
      Cries, consolable 4
      Inappropriate cry 3
      Persistent cry, grunts 2
      No response 1

TOTAL   **3-15**

Score of 8 or less: major injury with
significant mortality.

Fig. 7.8. Pediatric Glasgow Coma Scale.

present in 25% of cases. A petechial rash is present in meningococcemia. Other signs and symptoms are listed in the Fever and Septic Shock sections. In older children, fever, headache and photophobia are commonly present (see Chapter 6). **Treatment:** as for Septic Shock.

Shock, Seizures: See earlier sections

Alcohol/Drugs
As mentioned in Chapter 6, it is not appropriate to list the multitude of drugs and toxins that may cause a decreased level of consciousness. Alcohol

A. GLASGOW COMA SCALE (GCS): ⟶ GCS POINTS   CODE
    1. Eye opening:
           spontaneous 4   13-15 4
           to voice 3   9-12 3
           to pain 2   6-8 2
           none 1   4-5 1
                                                  3 0
    2. Verbal response (adult):
           oriented 5
           confused 4   ┌──────────────────────────┐
           inappropriate words 3   │ *TRAUMA SCORE IS SUM TOTAL* │
           incomprehensible sounds ...2   │   *OF CODES OF A, B, C*   │
           none 1   └──────────────────────────┘

    2a. Verbal response (pediatric):   *Score: 0 to 12*
           appropriate 5   *10 or less: major injury with*
           cries, consolable 4   *significant mortality*
           persistently irritable 3
           restless, agitated 2
           none 1

    3. Motor response
           obeys commands 6
           purposeful movements (pain) ...5
           withdraw (pain) 4
           flexion (pain) 3
           extension (pain) 2
           none 1

B. SYSTOLIC BLOOD PRESSURE (SBP):
       >89 4
       76-89 3
       50-75 2
       1-49 1
       0 0

C. RESPIRATORY RATE (RR):
       10-29 4
       >29 3
       6-9 2
       1-5 1
       0 0

Fig. 7.9. Pediatric Revised Trauma Score.

ingestion and intoxication is prevalent in the teenage population. Intentional and inadvertent drug ingestion is rising. **Treatment:** alcohol problems are discussed in Chapter 6. To prevent further absorption of ingested sedatives/hypnotics or opiates, activated charcoal (1 gm/kg) is administered by nasogastric tube (12-16 French). If no gag reflex is present, the patient is first intubated (rapid sequence intubation may be required—see Chapter 8).

Endocrine, Electrolytes: hypernatremia and hyponatremia (Na: <120 meq/l) cause weakness and decreased level of consciousness. Hypovolemic hypernatremia (from dehydration) is a common cause of hypernatremia and is treated with IV normal saline infusion. The fluid is changed to 0.5 NS when the urine output is 0.5 ml/kg/hr. Hyponatremia depends on the cause. If it is an excess of water, the treatment is water restriction. If it is because of Na loss, isotonic saline is administered.

Insulin: poorly controlled or new-onset diabetes mellitus is an important cause of decreased level of consciousness in pediatrics. Kussmaul breathing is frequently seen in diabetic ketoacidosis (see Chapter 6). Hypoglycemia (blood glucose <40 mg/dL) is rare, and is usually the result of poorly controlled insulin therapy or is idiopathic. **Treatment:** for hypoglycemia, glucose is administered as a 3 ml/kg IV bolus of D10W. For ketoacidosis, intravenous normal saline is infused at 20 ml/kg over 2 hours, and insulin is given at 0.1 unit/kg/hr. 40 meq of K+ is added to each liter of fluid.

Oxygen: see Chapter 8.

Uremia: see Chapter 6.

Practical Points

- First, the ABCs of resuscitation are followed (see Chapter 8).
- Pediatric arrests are usually respiratory.
- Children under 2 months of age with a temperature over 100.4 F (38°C) will usually require a septic workup, including lumbar puncture.
- Dehydration is a critical condition. Prolonged capillary refill and tachycardia are important signs of shock.
- Venous access in an infant is a problem for everyone. When necessary it must be done.

References

1. American Academy of Pediatrics/American College of Emergency Physicians. Advanced pediatric life support, 3rd Ed. Dallas, 1998.
2. American Academy of Pediatrics. Textbook of Pediatric advanced life support. Dallas: American Heart Association, 1997.
3. American College of Emergency Physicians (ACEP). Clinical policy for the initial approach to children under the age of 2 years presenting with fever. Ann Emerg Med 1993; 22:628.
4. American College of Surgeons: Pediatric trauma. In: Advanced Trauma Life Support. Chicago, 1997.
5. Apgar V. A proposal for a new method of evaluation of the newborn infant. Curr Research Anesth July-Aug 1953.
6. Apgar V et al. Evaluation of the newborn infant—Second report. JAMA 1958; 168:15.
7. Baker M. Evaluation and management of infants with fever. Ped Clin N Am 1999; 46:1061.
8. Baraff L. Management of fever without source in infants and children. Ann Em Med 2000; 36:602.
9. Barkin R, Rosen P. Emergency Pediatrics. St. Louis: CV Mosby Co., 1999.
10. Biehler J, Barnes B. Evaluation of vital signs. In: Henretig F et al. Textbook of Pediatric Emergency Procedures. Baltimore: Williams & Wilkins, 1997.

11. Bonadio W et al. Clinical characteristics of children with fever and transient neutropenia who experience serious bacterial infection. Ped Em Care 1989; 5:163.

12. Bulloch B, Ruddy M. Asthma update: Managing asthma in the pediatric emergency department. Em Med Rep 1998; 3:39.

13. Catalina P. Occult bacteremia in toddlers: Watch or treat? Em Med 1992; 24:67.

14. Catlin E et al. The APGAR score revisited: Influence of gestational age. J Pediatrics 1986; 109:5.

15. Cordle R, Relich N. Pediatrics: Upper respiratory emergencies. In: Tintinalli J et al. Emergency Medicine: A Comprehensive Study Guide. New York: McGraw-Hill, 2000.

16. Gjerris F. Head injuries in children—Special features. Acta Neurochi Suppl 1986; 36:155.

17. Gorelick M, Baker D. Epiglottitis in children, 1979 through 1992. Arch Pediatr Adolesc Med 1994; 148:47.

18. Haddock B et al. Axillary and rectal temperatures of full-term neonates: Are they different? Neonatal Network Aug, 1986.

19. Klassen T. Croup: A current perspective. Ped Clin N Am 1999; 46:1167.

20. Kuppermann N. Occult bacteremia in young febrile children. Ped Clin N Am 1999; 46:1073.

21. Landau L, Geelhoed G. Aerosolized steroids for croup. N Eng J Med 1994; 331:5.

22. Lewitt E et al. An evaluation of a plastic strip thermometer. JAMA 1982; 247:321.

23. Lieh-Lai M et al. Limitations of the Glasgow Coma Scale in predicting outcome in children with traumatic brain injury. J Ped 1992; 120:195.

24. Mace S. Issues in pediatric emergency medicine. Foresight 1999; 47:1.

25. McCarthy Paul. Infants with fever. NEJM 1993; 329:1494.

26. Morley C et al. Symptoms and signs in infants younger than 6 months of age correlated with the severity of their illness. Pediatrics 1991; 88:1119.

27. Nizet V et al. Fever in children. Ped Review 1994; 15:127.

28. Park M et al. Direct blood pressure measurements in brachial and femoral arteries in children. Circ Feb 1970:41.

29. Pidwell W et al. Accuracy of the temporal artery thermometer. Ann Em Med Suppl 2000; 36:5.

30. Ramenofsky M et al. Maximum survival in pediatric trauma: The ideal system. J Trauma 1984; 24:9.

31. Ramenofsky M et al. The predictive validity of the pediatric trauma score. J Trauma 1987; 27:7.

32. Rothrock S et al. Diagnosis of epiglottitis: Objective criteria for all ages. Ann Emerg Med 1990; 19:978.

33. Rothrock S, Perkin R. Stridor: A review, update, and current management recommendations. Em Med Rep 1996; 1:29.

34. Smith S, Strunk R. Acute asthma in the pediatric emergency department. Ped Clin N Am 1999; 46:1145.

35. Sofer S, Benkovich E. Severe infantile hypothermia: Short and long term outcome. Intens Care Med 2000; 26:88.

36. Supure J. Hyperpyrexia in children: Clinical implications. Ped Em Care 1987; 3:10.

37. Talan D et al. Analysis of emergency department management of suspected bacterial meningitis. Ann Em Med 1989; 18:856.

38. Thomson A et al. Validation of the Glasgow Meningococcal Septicemia Prognostic Score: A ten-year retrospective survey. Crit Care Med 1991; 19:26.

CHAPTER 8

Resuscitation

Adult Resuscitation

Resuscitation has become a ubiquitous term meaning not only cardiopulmonary arrest, but also prearrest conditions such as respiratory distress, stridor, coma, drug overdoses and some bradycardias and tachycardias. Excluding the hospitalized patient, the reason for the arrest is usually unknown and the only information is the rhythm on the monitor and tenuous vital signs. Thus, protocols involving therapies for specific conditions are often not helpful. In this book, the majority of prearrest conditions are discussed in the appropriate vital sign chapter. When vital signs are lost, basic resuscitation begins immediately (**Basic Life Support**) by any bystander ("lay rescuer" in the literature). When emergency medical services (EMS) arrives, Basic Life Support is supplemented with equipment and therapeutic modalities (**Advanced Life Support**. EMS and hospital personnel are "healthcare providers"). NOTE: 2000 BLS guidelines indicate that since the average person may have difficulty assessing a pulse, the pulse check is eliminated for the lay rescuer but applies to the healthcare provider.

Basic Life Support (BLS)

1. Establish unresponsiveness. Shake the person. "Are you okay?" If no response,
2. Activate EMS system, or if in hospital call for appropriate help.
3. **AIRWAY**. Open airway using jaw thrust or chin lift. If trauma suspected, stabilize neck, use jaw thrust.
4. **BREATHING**. Look, listen and feel for breathing. If inadequate or no breathing, 2 slow breaths mouth to mouth (nose closed), or use pocket mask or bag-valve mask, 12 breaths per minute.
5. **CIRCULATION**. If breathing, airway support as above. If no breathing (or no pulse by healthcare provider), begin chest compressions lower sternum, heel of hand (other hand on top), 2 inches deep, 100 per minute. 15 compressions and 2 ventilations for 1 or 2 rescuers (compression/ventilation ratio 15:2.

Advanced Life Support (ALS)

1. Assess responsiveness, if none,

Vital Signs and Resuscitation, by Joseph V. Stewart. ©2003 Landes Bioscience.

CPR/Rescue Breathing	Adult and Older Child	Child (Approximately 1–8 Years of Age)	Infant (Less Than 1 Year of Age)	Neonate and Newly Born
Establish unresponsiveness, activate EMS				
Open airway (head tilt–chin lift or jaw thrust)	Head tilt–chin lift (if trauma is present, use jaw thrust)	Head tilt–chin lift (if trauma is present, use jaw thrust)	Head tilt–chin lift (if trauma is present, use jaw thrust)	Head tilt–chin lift (if trauma, use jaw thrust)
Check for breathing: (Look, Listen, Feel) If victim breathing; place in recovery position. If victim not breathing; give 2 effective slow breaths				
Initial	2 effective breaths at 2 seconds per breath	2 effective breaths at 1 to 1½ seconds per breath	2 effective breaths at 1 to 1½ seconds per breath	2 effective breaths at approximately 1 second per breath
Subsequent	12 breaths/min (approximate)	20 breaths/min (approximate)	20 breaths/min (approximate)	30–60 breaths/min (approximate)
Foreign-body airway obstruction	Abdominal thrusts or back blows or chest thrusts	Abdominal thrusts or back blows or chest thrusts	Back blows or chest thrusts (no abdominal thrusts)	Back blows or chest thrusts (no abdominal thrusts)

Fig. 8.1. Summary of Basic Life Support (BLS) Protocols (continued below). Reprinted with permission from: Guidelines for 2000 for Cardiopulmonary Resuscitation and Emergency Cardiovascular Care, American Heart Association.

	Pulse check (healthcare providers)* Carotid	(Healthcare providers)* Carotid	(Healthcare providers)* Brachial	(Healthcare providers)* Umbilical
Signs of Circulation: Check for normal breathing, coughing or movement, pulse.* If signs of circulation present, provide airway and breathing support. If signs of circulation absent, begin chest compressions interposed with breaths				
Compression landmarks	Lower half of sternum	Lower half of sternum	Lower half of sternum (1 finger width below intermammary line)	Lower half of sternum (1 finger width below intermammary line)
Compression method	Heel of 1 hand, other hand on top	Heel of 1 hand	Two thumbs–encircling hands for 2-rescuer healthcare provider or 2 fingers	Two thumbs–encircling hands for 2-rescuer healthcare provider or 2 fingers
Compression depth	Approximately 1½ to 2 in	Approximately ⅓ to ½ the depth of the chest (1 to 1½ in)	Approximately ⅓ to ½ the depth of the chest (½ to 1 in)	Approximately ⅓ to ½ the depth of the chest
Compression rate	Approximately 100/min	Approximately 100/min	At least 100/min	Approximately 120 events/min (90 compressions/30 breaths)
Compression/ventilation ratio	15:2 (1 or 2 rescuers, unprotected airway) 5:1 (2 rescuers, protected airway)	5:1 (1 or 2 rescuers)	5:1 (1 or 2 rescuers)	3:1 (1 or 2 rescuers)

*Pulse check is performed as one of the "signs of circulation" assessed by healthcare providers. Lay rescuers check for other signs of circulation but do not check pulse.

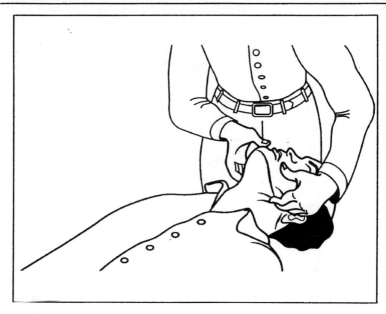

Fig. 8.2. Jaw Thrust.

2. Outside hospital—activate EMS system.
   Inside hospital—call for defibrillator, or if on monitor and defribillator present and V-fib or pulseless V-tach defibrillate x 3 (200, 300 , 360 joules).
3. Open **AIRWAY** and assess **BREATHING**. If none, 2 slow breaths (in trauma, maintain cervical spine immobilization).
4. Assess **PULSE**. If present, mouth-to-mouth or bag-valve-mask or intubation at 12 breaths per minute (plus needle thoracentesis or chest tube if needed). If no pulse,
5. **CPR:** chest compressions 100 per minute. Compression/ventilation for 1 to 2 rescuers (15:2), with intubated patient 2 providers (5:1).
6. When **DEFIBRILLATOR** arrives and shows VF/pVT, defibrillate x 3. If another rhythm, continue CPR, intubate, IV access, determine rhythm and therapy.
7. **DISABILITY:** if adequate pulse and respirations, but comatose, use **Coma Protocol** (Fig. 6.4).

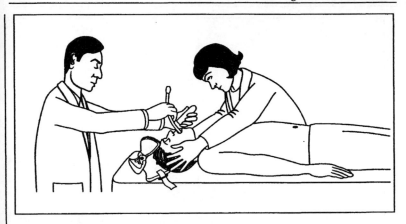

Fig. 8.3. Intubation with In-line Immobilization.

1. AIRWAY:

Airway control is the critical first step in resuscitation. The head is tilted and either the jaw thrust or chin lift used to open the airway. Protection of the airway may require endotracheal intubation, often needed with respiratory distress and hypoventilation from any cause. It is also needed to prevent aspiration in the comatose patient or the patient with decreased level of consciousness and no gag reflex. The patient is ventilated with 100% oxygen using two assistants and a bag-valve-mask. Insertion of an endotracheal tube is performed orally (**orotracheal intubation**) or nasally (**nasotracheal intubation**). Nasotracheal intubation is performed only on the breathing patient. It is useful in the immobilized patient, when the patient has clenched teeth or when cooperation is needed (i.e., the asthmatic or the patient with pulmonary edema). It is contraindicated with maxillofacial injuries or a basilar skull fracture.

Trauma

In the trauma patient, the C-spine is protected with a cervical collar and the jaw thrust is used. The oral cavity is inspected for foreign bodies, vomitus, broken teeth and suctioned using a hard-tipped suction catheter. Examination of the neck and **intubation in the trauma patient** requires **in-line immobilization** by an assistant in order to stabilize the neck when the cervical collar is removed (Fig. 8.3).

Rapid Sequence Intubation

Intubation is sometimes necessary for an awake patient. The criterion is deterioration of vital signs so that the airway will be unprotected. Examples: head injury, a combative or agitated patient, a patient with clenched teeth,

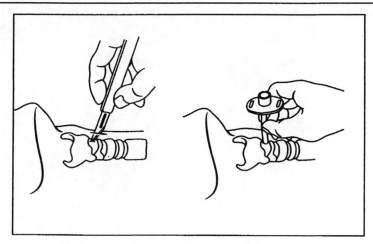

Fig. 8.4. Cricothyrotomy.

respiratory failure, impending cardiovascular collapse and a diminishing level of consciousness from any cause. Rapid sequence intubation (RSI) is performed using the following or a similar protocol:

1. Oxygen by nonrebreather mask for 5 minutes.
2. Induction with IV thiopental (Pentothal) 4 mg/kg or etomidate (Amidate) 0.3 mg/kg. Lidocaine 1 mg/kg IV is added to blunt the rise in intracranial pressure in head injuries.
3. An assistant applies pressure to the cricoid cartilage to occlude the esophagus and prevent esophageal reflux and aspiration (Sellick maneuver).
4. Paralysis with IV succinylcholine (Anectine) 1.5 mg/kg, or rocuronium (Zemuron) 1 mg/kg.
5. The tube is passed through the vocal cords (intubation).
6. Cricoid pressure is released.
7. Bag, listen for breath sounds, pulse ox/CO2 indicator.
8. X-ray for tube placement.

Surgical Airway

If intubation is not possible, a tracheostomy tube with an inflatable cuff (i.e., #6 Shiley), or a #6 or 7 endotracheal tube, is inserted through the cricothyroid membrane between the thyroid ("Adam's apple") and cricoid cartilages (**cricothyrotomy**). The patient is bagged through the ET or tracheostomy tube. If a tracheostomy tube is not available, a 14g angiocath may be inserted through the cricothyroid membrane (**needle cricothyrotomy**) while preparing for a surgical airway (Fig. 8.4).

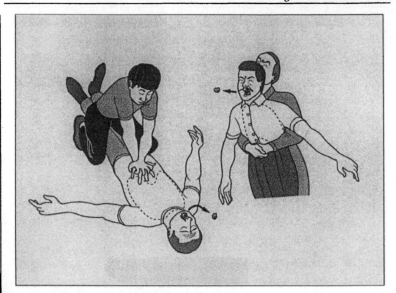

Fig. 8.5. Abdominal Thrust (Heimlich Maneuver).

Choking

Choking from a foreign-body does not occur in the emergency department, but often takes place in a restaurant ("cafe coronary"). The person is drinking, the throat muscles are relaxed, he is talking while eating and inhales at the same time he is swallowing. The piece of food, usually meat, impacts against the vocal cords, obstructing breathing. The person grasps his neck with the thumb and fingers, a universally recognized sign of airway obstruction. The first thing to do is to ask if he can speak. If he can, he is not obstructed. If he cannot, he is. Partial obstruction may occur, in which case coughing and stridor are present, but the person is exchanging air. Treatment for obstruction is as follows (Fig. 8.5):

1. The **abdominal thrust (Heimlich maneuver)** creates intrathoracic pressure and will expel almost any foreign body. The thumb side of the fist is placed against the person's abdomen, slightly above the navel. The fist is grasped with the other hand and pressed upward quickly and forcefully as many times as needed to dislodge the object.

2. If the person becomes unconscious, he is placed on his back on the floor. With one hand on top of the other, abdominal thrusts are performed by the rescuer with the palms of the hands pushing upward.

3. Since hypoxic brain damage occurs in 7-10 minutes, after several failed attempts at abdominal thrusts the rescuer should open the mouth and perform finger sweeps. After this, rescue breathing should be attempted.

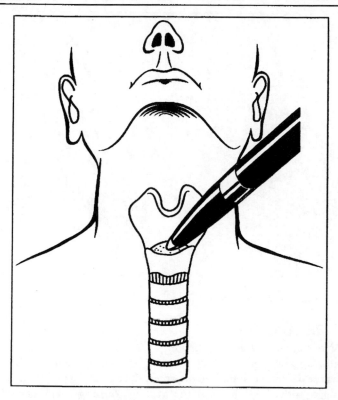

Fig. 8.6. Emergency Cricothyrotomy.

4. If attempts at abdominal thrusts, finger sweeps and rescue breathing are unsuccessful and the 7 minute point is approaching, an **emergency cricothyrotomy** may be life-saving: thrust anything, such as a knife or ballpoint pen, into the indentation between the thyroid and cricoid cartilages (cricothyroid membrane). The force must be substantial or the trachea will not be penetrated. Remove the inner pen piece and keep the barrel in place, or keep the space open with two keys, or 2 to 3 straws. Stay with the person until paramedics arrive.

5. The Heimlich maneuver may be performed on oneself, as in 1.

Treatment for partial obstruction: remain with the person until he is transported to the nearest emergency facility, since complete obstruction may occur at any time.

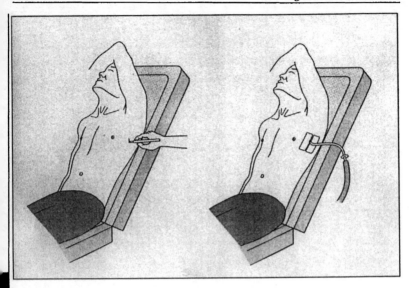

Fig. 8.7. Chest Tube Placement.

2. BREATHING:

After securing the airway, the lungs are auscultated for bilateral breath sounds. Breathing patients receive 100% oxygen by nonrebreather mask. Comatose patients are intubated to protect the airway. Nonbreathing patients are bagged with a bag-valve-mask (BVM) at 100% oxygen and are intubated.

Trauma

If signs of **tension pneumothorax** or hemothorax are present or evolving (chest pain, dyspnea, decreased breath sounds on affected side, tracheal deviation, jugular venous distention), a 14g needle or angiocath is inserted in the 2nd interspace at the mid-clavicular line (**needle thoracentesis**) while preparing for **chest tube** (thoracostomy tube) placement, before a chest x-ray is taken (see Fig. 8.18). A 36F chest tube is inserted in the 5th intercostal space at the midaxillary line over the top of the rib (to avoid vessels) and connected to an underwater seal apparatus (Fig. 8.7).

Paradoxic motion of the chest wall from moving rib segments (flail chest) may require intubation. An open wound of the chest wall (open pneumothorax) requires a sterile occlusive dressing taped on three sides, providing a flutter-type valve effect, followed by insertion of a chest tube.

Respiratory Failure

Respiratory failure is seen in asthma, congestive heart failure, COPD, trauma (i.e., pulmonary contusion, pneumo-hemothorax) and occasionally pneumonia. Signs of hypoxia are dyspnea, tachypnea, tachycardia, restlessness, gasping respirations and use of accessory ventilatory muscles. Lethargy and confusion are seen with **hypercapnia** (see Chapter 6, Oxygen). ABGs show a PO_2 <50 mmHg and/or a PCO_2 >50 mmHg, implying impending respiratory failure, although patients with COPD may normally carry a PCO_2 of 60-70 mmHg. A rectal temperature is taken, since the person is mouth-breathing. **Treatment:** endotracheal intubation is usually required, although some cases may respond to continuous positive airway pressure (CPAP). Initial settings on a volume-cycled respirator are: oxygen 100%, tidal volume 15 cc/kg, respiratory rate 16.

3. CIRCULATION:

Hemorrhage is controlled by pressure. Blood loss is treated with 2 large bore IVs, 2 liters of normal saline and type-specific or O-neg packed red blood cells (RBCs). Treatments for hypovolemic, cardiogenic (myocardial infarction, aortic aneurysms, cardiac tamponade), septic and anaphylactic shock, as well as hypertensive emergencies requiring resuscitation, are discussed in Chapter 5.

Pulseless Rhythms

Pulseless rhythms are **ventricular fibrillation, pulseless ventricular tachycardia, pulseless electrical activity** and **asystole,** the latter being often a terminal event. Other rhythms (bradycardias, tachycardias, etc) are discussed in Chapter 3. One must be careful to **examine the patient** and not the monitor. The monitor may show a normal sinus rhythm but the patient may be apneic or pulseless. Conversely, the monitor may show a chaotic rhythm or straight-line, but if the patient is alert and conversant, a lead is off.

In **ventricular fibrillation (VF)**, the electrical activity of the heart is chaotic and no heart-beat is present. **Pulseless ventricular tachycardia (pVT)** shows VT but without a pulse and is treated as VF (one must be careful not to confuse this with ventricular tachycardia with a pulse—see Chapter 3). **Treatment:** CPR is begun and the patient is defibrillated as soon as possible 3 times in succession (200, 300, 360 J). If unsuccessful the patient is intubated and CPR is continued. Epinephrine is given 1 mg q 5 min. Vasopressin 40 units may be given as one dose (vasopressin at this dosage is a vasoconstrictor, and is frequently used in Europe). It has been shown that antiarrhythmic agents possess minimal efficacy in VF/pVT. The usual protocol is to give the drug, followed by defibrillation. However, it is acceptable to give the agent, followed by three shocks. Agents used, in order of preference, are:

1. Amiodarone 300 mg IV push. A second dose of 150 mg may be given,

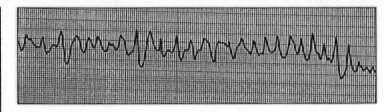

Fig. 8.8. Ventricular Fibrillation. Reprinted with permission from: Merck, Sharp & Dohm, Division of Merck & Co., Inc.

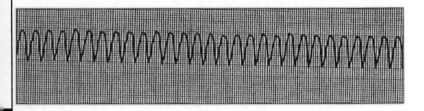

Fig. 8.9. Pulseless Ventricular Tachycardia. Reprinted with permission from: Merck, Sharp & Dohm, Division of Merck & Co., Inc.

2. Lidocaine 1 mg/kg IV push, and repeat in 5 minutes to a total of 3 mg/kg. Defibrillation may be after each agent, or after each minute of CPR.

In **pulseless electrical activity (PEA)**, the monitor shows a rhythm, but the patient has no heart beat—the electrical activity is inadequate to stimulate contraction of the heart muscle, or the contraction is so weak as to be negligible. It is seen in several circumstances, the more common being hypovolemia and massive acute myocardial infarction. **Treatment:** this is a situation in which the patient may be mistakenly assumed to have a pulse. Always check for a pulse. Unfortunately, the reason for this lethal condition is often not known. CPR is begun, intubation is performed, IV access is obtained and epinephrine 1 mg IV push is given every 5 minutes. If electrical bradycardia is present, atropine 1mg IV is given every 5 minutes to a total of 0.04mg/kg. Because this condition is reversible in some circumstances, as a last resort bicarb 1 meq/kg and a 200 cc bolus x 2 of normal saline may be tried.

Asystole, or a straight line on the monitor, is treated with CPR, transcutaneous pacing if the rhythm occurred suddenly, epinephrine 1 mg IV every 5 min and atropine 1 mg IV q 5 min (total 0.04 mg/kg). This is often a terminal nonrhythm, indicating death.

Some bradycardias and tachycardias may require resuscitative measures (see Chapter 3).

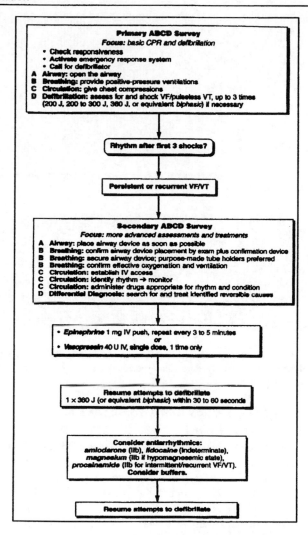

Fig. 8.10. Ventricular Fibrillaton/Pulseless Ventricular Tachycardia Algorithm. Reprinted with permission from: Guidelines for 2000 for Cardiopulmonary Resuscitation and Emergency Cardiovascular Care, American Heart Association.

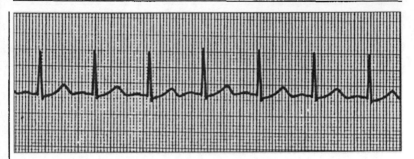

Fig. 8.11. Pulseless Electrical Activity (PEA). Reprinted with permission from: Merck, Sharp & Dohm, Division of Merck & Co., Inc.

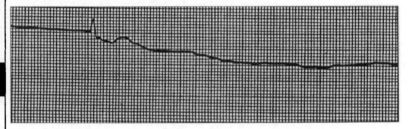

Fig. 8.12. Asystole. Reprinted with permission from: Guidelines for 2000 for Cardiopulmonary Resuscitation and Emergency Cardiovascular Care, American Heart Association.

4. DEFIBRILLATION and DISABILITY (Level of Consciousness)

In the hospital and emergency medical services (EMS) settings, "D" for Defibrillation is added to the ABCs. When a monitor or "quick look" paddles show ventricular fibrillation or pulseless ventricular tachycardia, the patient is defibrillated immediately as per the above protocol. In the trauma and other settings, "D" also represents "Disability", or level of consciousness. Assessment and management for level of consciousness and therapy for increased intracranial pressure is described in Chapter 6 (see Figs. 6.4, 6.5).

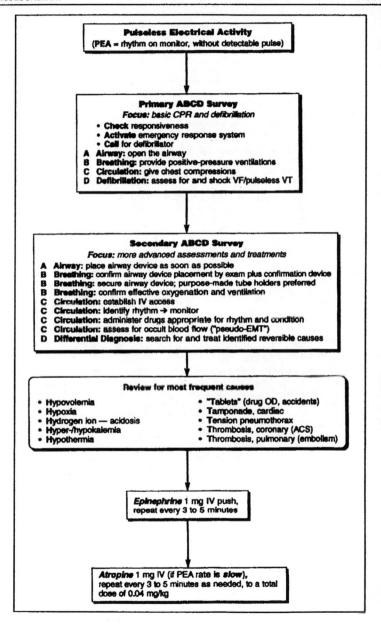

Pulseless Electrical Activity
(PEA = rhythm on monitor, without detectable pulse)

Primary ABCD Survey
Focus: basic CPR and defibrillation
- **Check** responsiveness
- **Activate** emergency response system
- **Call** for defibrillator
A **Airway:** open the airway
B **Breathing:** provide positive-pressure ventilations
C **Circulation:** give chest compressions
D **Defibrillation:** assess for and shock VF/pulseless VT

Secondary ABCD Survey
Focus: more advanced assessments and treatments
A **Airway:** place airway device as soon as possible
B **Breathing:** confirm airway device placement by exam plus confirmation device
B **Breathing:** secure airway device; purpose-made tube holders preferred
B **Breathing:** confirm effective oxygenation and ventilation
C **Circulation:** establish IV access
C **Circulation:** identify rhythm → monitor
C **Circulation:** administer drugs appropriate for rhythm and condition
C **Circulation:** assess for occult blood flow ("pseudo-EMT")
D **Differential Diagnosis:** search for and treat identified reversible causes

Review for most frequent causes
- Hypovolemia
- Hypoxia
- Hydrogen ion — acidosis
- Hyper-/hypokalemia
- Hypothermia
- "Tablets" (drug OD, accidents)
- Tamponade, cardiac
- Tension pneumothorax
- Thrombosis, coronary (ACS)
- Thrombosis, pulmonary (embolism)

Epinephrine 1 mg IV push,
repeat every 3 to 5 minutes

Atropine 1 mg IV (if PEA rate is *slow*),
repeat every 3 to 5 minutes as needed, to a total
dose of 0.04 mg/kg

Fig. 8.13. Pulseless Electrical Activity Algorithm. Reprinted with permission from: Guidelines for 2000 for Cardiopulmonary Resuscitation and Emergency Cardiovascular Care, American Heart Association.

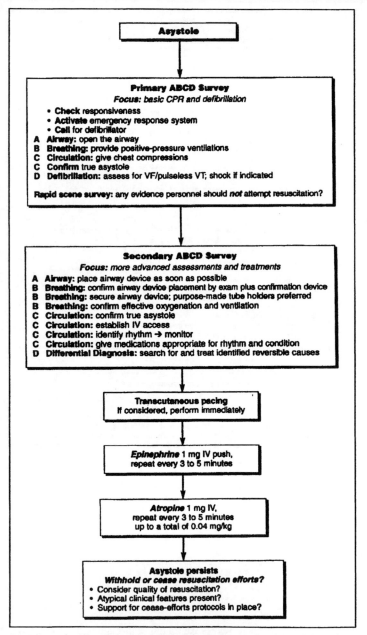

Fig. 8.14. Asystole Algorithm. Reprinted with permission from: Guidelines for 2000 for Cardiopulmonary Resuscitation and Emergency Cardiovascular Care, American Heart Association.

Primary Survey/Resuscitation:

Airway/C-spine:
open airway (jaw thrust, chin lift), oxygen, secure airway (intubation, cricothyrotomy), pulse oximetry, clear c-spine as soon as possible

Breathing:
bag-valve-mask, ventilator, needle decompression, chest tube placement, treat pulmonary edema

Circulation:
cardiac monitor, immediate IV access
pulse --- treat dysrhythmias (Ch 3, 8)
shock --- treat (Ch 5, 8)

Disability: neurological status
if decreased level of consciousness: chemstrip glucose (if unavailable, D-50 1 amp, peds: D-25 4 cc/kg), thiamine 100 mg, naloxone 2 mg IV (peds: 0.1 mg/kg) (Ch 6), Defibrillation

Secondary Survey:
Complete physical exam
History

Fig. 8.15. Resuscitation Protocol.

Pediatric Resuscitation

Pediatric Basic Life Support

1. Establish unresponsiveness. Tap the child and speak loudly.
2. Call for appropriate help.
3. **AIRWAY.** Open airway using jaw thrust or chin lift (if trauma suspected, stabilize neck, use jaw thrust).
4. **BREATHING.** Look, listen and feel for breathing. If no breathing, 2 slow breaths mouth to mouth (nose closed). In infants (1 year and less) rescue mouth over nose and mouth. Rescue breathing at 20 breaths per minute.
5. **CIRCULATION.** If no breathing, begin chest compressions alternating with ventilations, 2 or 3 fingers lower sternum (heel of one hand with larger children), 1/2 to 1 inch deep, 100 per minute (for healthcare providers: pulse check, if pulse <60 begin chest compressions). Compression/ventilation ratio 5:1 1-2 rescuers.

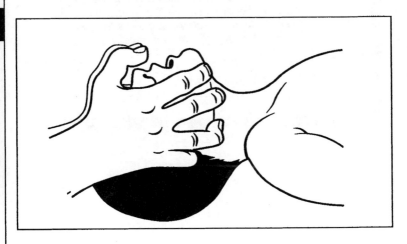

Fig. 8.16. Peds Jaw Thrust.

Pediatric Advanced Life Support

1. AIRWAY

As in the adult, the head is tilted and either the jaw thrust or chin lift used to open the airway. In the child requiring intubation, the patient is first ventilated by bag-valve-mask with 100% oxygen. If the child is slightly breathing, gentle positive-pressure should be carefully timed with voluntary respirations. Unlike the adult where two assistants are required to adequately bag the patient, one assistant is often sufficient. In the infant, the jaw is supported with the base of the middle and 4th fingers. In older children, the fingertips of the 3rd, 4th and 5th fingers are placed on the ramus of the mandible to hold the jaw forward and extend the head. Endotracheal intubation is always via the orotracheal route (nasotracheal intubation is not performed in children). Rapid sequence intubation (RSI) is accomplished as in the adult (see earlier section).

Trauma

In major trauma, the c-spine is immobilized and the jaw thrust is performed. The oral cavity is inspected for foreign bodies, vomitus, broken teeth and suctioned using a hard-tipped suction catheter of appropriate size. Not only must the C-spine be cleared but the child must be cleared neurologically. If the history and physical exam indicate a possible spinal cord injury (spinal cord injury without radiographic abnormality—SCIWORA) the C-collar is left on and the patient is cleared by the neurosurgeon.

Airway Obstruction

Diagnosing a foreign body in the airway may pose a difficult problem unless complete obstruction occurs. Offenders are nuts, toy parts, round candies and aluminum "pop-tops". Complete obstruction in an infant is treated by a variation of the Heimlich maneuver: the infant is held prone in the left hand and forearm and 5 back blows are delivered between the shoulder blades with the heel of the right hand. Then the infant is turned over, with the head lower than the body, and 5 quick chest thrusts are delivered on the lower third of the sternum. The mouth is opened and, if visualized, the foreign body is removed. The finger sweep and rescue breathing are performed. Nearly all larger foreign bodies are captured at this point. Smaller foreign bodies will be moved into a mainstem bronchus. In the child over 1-2 years, the Heimlich maneuver is similar to the adult.

If the patient can not be adequately bagged or intubated, a needle cricothyrotomy is performed by inserting a 14 or 16g angiocath through the cricothyroid membrane. The needle is removed, the cannula is secured and attached to oxygen tubing using a "Y" connector, at 20 breaths per minute: 1 second inhalation, 2 seconds exhalation. A surgical cricothyrotomy is not performed in children less than 9 years old (Figs. 8.17, 8.18).

2. BREATHING

As in the adult, the lungs are auscultated for equal breath sounds. Breathing children receive 100% oxygen by nonrebreather mask. Comatose patients are intubated to protect the airway. Nonbreathing children are bagged with a bag-valve-mask and 100% oxygen and are intubated.

Trauma

In trauma, if signs of tension pneumothorax are present (respiratory distress, distended neck veins, tracheal deviation), needle decompression is

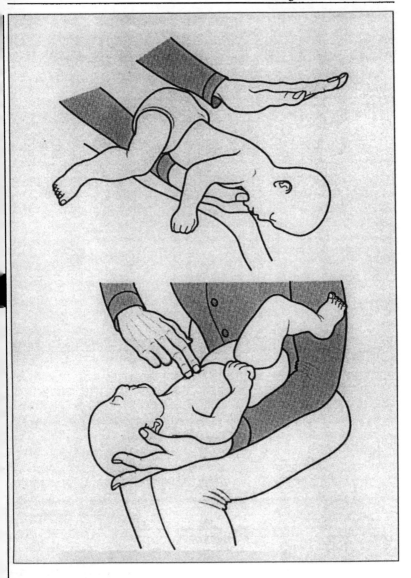

Fig. 8.17. Infant Foreign-Body Airway Obstruction.

accomplished through the 2nd intercostal space above the 3rd rib at the midclavicular line, followed by chest tube placement at the 5th interspace anterior to the mid-axillary line. An open pneumothorax is treated with an occlusive dressing and a chest tube as described for the adult.

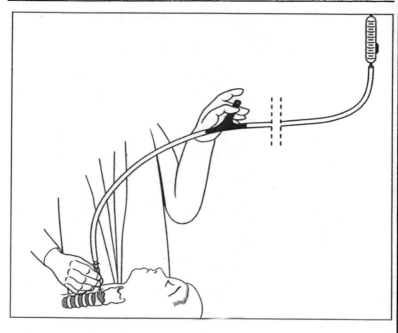

Fig. 8.18. Peds Needle Cricothyrotomy.

Respiratory Failure

In respiratory failure, oxygen by mask should be administered to all seriously ill or injured children. A nasopharyngeal airway is placed in the conscious patient. Respiratory failure should be anticipated when the following signs are present:

1. Increased respiratory effort. Tachypnea is the first sign of respiratory distress in infants. Other signs are restlessness, use of accessory muscles with nasal flaring, inspiratory/intercostal/sternal retractions and tachycardia. Stridor is an inspiratory high-pitched sound of upper airway obstruction. Wheezing may be present. Grunting is caused by premature closure of the glottis during early expiration in an attempt to increase airway pressure.
2. Diminished level of consciousness or response to pain.
3. Poor skeletal muscle tone.
4. Cyanosis is a late sign.

Treatment of Respiratory Failure:

1. Open airway.
2. 100% oxygen by mask.
3. If the patient is not moving air, begin bag-mask ventilations with small volumes and prepare for endotracheal intubation.

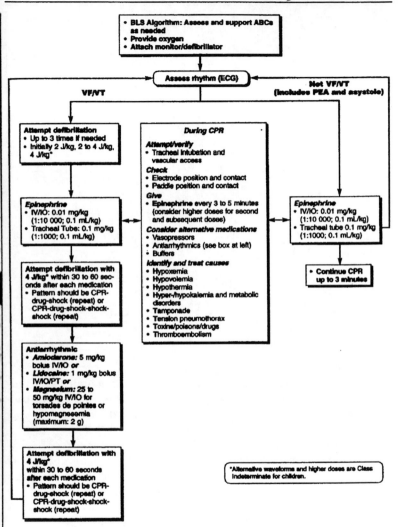

Fig. 8.19. Peds Ventricular Fibrillation/Asystole/PEA Algorithm. Reprinted with permission from: Guidelines for 2000 for Cardiopulmonary Resuscitation and Emergency Cardiovascular Care, American Heart Association.

3. CIRCULATION

In a child with no pulse, chest compressions are begun and an intravenous line is secured. If intravenous access cannot be obtained within 1-2 minutes, **intraosseous access** should be performed at the proximal medial tibia. The next step depends on the reason for the arrest. In adults it is often cardiac, but in children it is usually respiratory or traumatic. In the trauma

Component	+2	+1	-1
Size	>20 kg	10-20 kg	≤10 kg
Airway	Normal	Maintainable	Unmaintainable
CNS	Awake	Obtunded	Coma
Systolic BP	≥90 mm Hg	50-90 mm Hg	≤50 mm Hg
Open wound	None	Minor	Major
Skeletal injuries	None	Closed fracture	Open or multiple fractures

Score: -6 to +12
8 or less: major injury with
significant mortality

Fig. 8.20. Pediatric Trauma Score (PTS).

setting, hemorrhage is identified and controlled. Hemorrhagic shock is treated with normal saline boluses (20 cc/kg x 3) and type-specific or O-neg packed red cells (10 cc/kg) as indicated.

Ventricular fibrillation/pulseless ventricular tachycardia, asystole and **pulseless electrical activity** are rare occurrences in pediatrics. In the hospital, a call is made for quick-look paddles/monitor, with appropriate therapy depending on the arrhythmia (Fig. 8.19, see also Figs. 8.8, 8.9, 8.11, 8.12).

4. DISABILITY (Level of Consciousness)

"D" for Disability is similar to the adult and represents Level of Consciousness. As mentioned in Chapter 7, the Glasgow Coma Scale has been modified for infants ages 0-23 months and children ages 2-5 years (Pediatric Glasgow Coma Scale—PGCS). Since the total scores are the same as for the adult, intubation is still required for a score of 8 or less. As in the adult, AVPU is sometimes substituted for the PGCS in the field (see Chapter 6). The PGCS is now an integral part of the Revised Trauma Score (RTS) (see Figs. 7.8 and 7.9). A separate **Pediatric Trauma Score (PTS)** has also been developed that does not use the GCS. Children with an RTS of less than 12 or a PTS of less than 8 are at increased risk for morbidity and should be evaluated at a trauma center. Management of coma and therapy for increased intracranial pressure is described in Chapter 6 (Figs. 6.4, 6.5) (Fig. 8.20).

Neonatal Resuscitation

During delivery, as soon as the head is delivered, the mouth and the nose are suctioned. If meconium is present, intubation is performed and suction is applied through the endotracheal tube. The newborn is dried, warmed, positioned supine, suctioned and tactile stimulation is applied. For low Apgars, oxygen is administrated (see Chapter 7). If the heart rate is less than

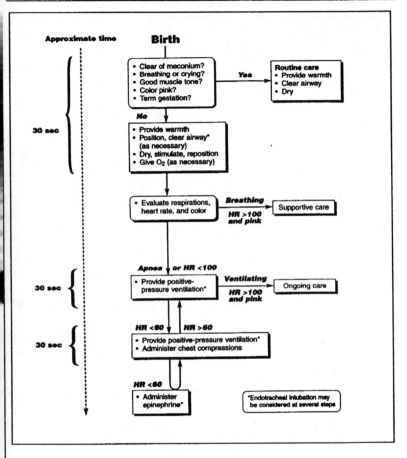

Fig. 8.21. Neonatal Resuscitation Algorithm. Reprinted with permission from: Guidelines for 2000 for Cardiopulmonary Resuscitation and Emergency Cardiovascular Care, American Heart Association.

100 beats per minute, the newborn is ventilated with a bag-valve-mask at 40 breaths per minute. If bag-mask ventilation cannot raise the heart-rate above 100, intubation is required. Chest compressions (90 per minute) and ventilations (30 per min) are performed if the heart rate is less than 60. If the heart-rate remains below 60 after intubation, epinephrine 0.01 mg/kg (1:10,000) is administered by way of the umbilical vein. Neonatal resuscitation is sometimes necessary because of drug ingestion, particularly crack/cocaine, by the mother. If the history indicates drug abuse by the mother, **naloxone** (0.1 mg/kg) IV is given (Fig. 8.21).

Special Resuscitation Cases

A multitude of drugs, toxins and circumstances may cause near-arrest or arrest. Space does not permit a discussion of them. Hypothermia and hyperthermia are described in Chapter 2. Some toxins, including carbon monoxide, are discussed in Chapter 6.

Electrical and Lightning Injuries

Arrest may occur from high voltage direct current (> 1000 volts). Do not attempt removal until current is shut off. With low voltage exposure, such as 120 volts of alternating current (AC) from a hair dryer or radio, for example, removal can be accomplished with cloth, rubber or wood. High voltage (DC) contact usually causes asystole; low voltage (AC) usually causes ventricular fibrillation. A lightning strike may carry up to a billion volts of DC energy and may strike directly, obliquely or "splash" from contact as it strikes the ground or at an object and spreads out to involve the victim. **Treatment:** the ABCs are assessed and CPR is begun, followed by intubation and IV access. Defibrillation or treatment for asystole is performed as required.

Hyperkalemia

Although hypokalemia and hyperkalemia may both cause arrest (in hypokalemia the heart stops in systole; in hyperkalemia it stops in diastole), hyperkalemia is the more potentially fatal condition, sometimes seen in the dialysis patient (i.e., K+ level above 8 mEq/L). Less common causes are crush injuries, burns, acidosis and some drugs (i.e., digoxin, ACE-inhibitors). It is often impossible to learn the reason for the arrest other than a fragmentary history (i.e., dialysis). **Treatment:** the ABC's are assessed and CPR is begun, followed by intubation and IV access. Defibrillation or treatment for asystole is instituted as needed. If the cause is known, 10 cc of a 10% solution of calcium chloride is administered. Calcium stabilizes the cardiac cell membrane, preventing the continued adverse effect of the increased K+. If the resuscitation is successful, glucose 2 amps + regular insulin 10 units IV push plus bicarb 1 amp will lower the K+ by moving it into the cells, thus decreasing the level in the bloodstream.

Pregnancy

The main causes of cardiac arrest in pregnancy are pulmonary embolism, trauma (high homicide), hemorrhage and pre-existing cardiac disease. Pulmonary embolism is discussed in Chapter 4. Hemorrhagic shock is discussed in Chapter 5. **Treatment:** the cause is often not known (other than trauma). Usual therapies such as heparin for pulmonary embolism will probably not be effective. The patient in late pregnancy is turned to the left and a roll or Cardiff wedge is placed under the right flank and hip for CPR. Appropriate therapy for cardiac arrhythmias and blood loss (IV NS and O-neg packed cells) is instituted. If pulmonary embolism seems likely, thrombolytics may be tried. Emergency cesarean section is performed if 5 minutes have elapsed, the fetus is viable (>20 weeks) and therapy has not been successful. The results are poor if the C-section is delayed for up to 15 minutes.

Submersion (Near-Drowning)

NOTE: the 2000 International Liaison Committee on Resuscitation (ILCOR) recommends using the word "submersion" and discarding "near-drowning" ("drowning" is considered submersion with death at the scene). Teenagers and toddlers are frequently affected. Drugs, alcohol (60% of teenage drownings), cervical spine injury, seizure disorders, mental retardation, child abuse and suicide are contributory. Death/arrest is from laryngospasm and asphyxia. Hypothermia is frequently present. **Treatment:** CPR is begun after removal from the water, with C-spine protection if a head-injury is suspected. On EMS arrival, the ABC's are assessed (including C-spine immobilization) and CPR continued as needed. Supplemental oxygen by mask is administered to all patients, with fast transport. In the emergency department IV access is secured, and intubation is performed as needed. Appropriate therapy for cardiac arrhythmias is instituted. If hypothermia is present, rewarming measures are begun and resuscitative measures are continued until the temperature is 90°F (see Chapter 2).

References

1. American College of Surgeons. Advanced trauma life support. Chicago, 1997.
2. American Heart Association and the International Liaison Committee on Resuscitation (ILCOR). Guidelines 2000 for cardiopulmonary resuscitation and emergency cardiovascular care. Baltimore: Lippincott, Williams & Wilkins, 2000.
3. American Academy of Pediatrics/American Heart Association. Pediatric advanced life support. Dallas, 1997.
4. Barkin R, Rosen P. Emergency pediatrics: A guide to ambulatory care. St. Louis: CV Mosby, 1999.
5. Berg R et al. Bystander cardiopulmonary resuscitation: Is ventilation necessary? Circulation 1993; 88:1907.
6. Bessen H. Hypothermia. In: Tintinalli J et al. Emergency Medicine: A Comprehensive Study Guide. New York: McGraw-Hill, 2000.
7. Cobb L et al. Influence of cardiopulmonary resuscitation prior to defibrillation in patients with out-of-hospital ventricular fibrillation. JAMA 1999; 281:1182.
8. Colquhoun M et al. ABC of Resuscitation. London: BMJ Books, 1999.
9. Cummins R, Hazinski M. Cardiopulmonary resuscitation techniques and instruction: When does evidence justify revision? Ann Em Med 1999; 34:780.
10. Datner E, Promes S. Resuscitation issues in pregnancy. In: Tintinalli J et al. Emergency Medicine: A Comprehensive Study Guide. New York: McGraw-Hill, 2000.
11. Dayton L. Secrets of a bolt from the blue. New Scientist 1993; 140:16.
12. De Maio V et al. Cardiac arrest witnessed by emergency medical services personnel: Descriptive epidemiology, prodromal symptoms, and predictors of survival. Ann Em Med 2000; 35:138.
13. de Vos R et al. In-hospital cardiopulmonary resuscitation: Prearrest morbidity and outcome. Arch Int Med 1999; 159:845.

14. Ewy G. Cardiopulmonary resuscitation—Strengthening the links in the chain of survival. New Engl J Med 2000; 342:1599.
15. Finer N et al. Cardiopulmonary resuscitation in the very low birth weight infant: The Vermont Oxford Network experience. Pediatrics 1999; 104:428.
16. Goodlin S et al. Factors associated with use of cardiopulmonary resuscitation in seriously ill hospitalized adults. JAMA 1999; 282:2333.
17. Guohua L et al. Cardiopulmonary resuscitation in pediatric trauma patients: Survival and functional outcome. J Trauma 1999; 47:1.
18. Hallstrom A et al. Cardiopulmonary resuscitation by chest compression alone or with mouth-to-mouth ventilation. N Engl J Med 2000; 342:1546.
19. Hals G, Crump T. The pregnant patient: Guidelines for management of common life-threatening medical disorders in the emergency department. Em Med Reports 2000; 21:53.
20. Jiva T. Critical care of pregnant women, Part 1: Pulmonary edema, ARDS, thromboembolism. J Crit Illness 2000; 15:316.
21. Kirkegaard-Nielsen H et al. Rapid tracheal intubation with rocuronium. Anesthesiology 1999; 91:131.
22. Krumholz H et al. Resuscitation preferences for heart-failure patients likely to change. Circulation 1998; 98:648.
23. Kudenchuk P. Amiodarone for resuscitation after out-of-hospital cardiac arrest due to ventricular fibrillation. N Engl J Med 1999; 341:871.
24. Landwirth J. Ethical issues in pediatric and neonatal resuscitation. Ann Emerg Med 1993; 22:502.
25. Losek J. Hypoglycemia and the ABC's (sugar) of pediatric resuscitation. Ann Emerg Med 2000; 35:43.
26. Markovchick V, Pons P. Emergency Medicine Secrets. Philadelphia: Hanley & Belfus, 1999.
27. Noe M et al. Mechanical ventilation may not be essential for initial cardiopulmonary resuscitation. Chest 1995; 108:821.
28. Patterson M. Resuscitation update for the pediatrician. Ped Clin N Am 1999; 46:1285.
29. Reed W. Near-drowning: Life-saving steps. Phys and SpoRTSmed 1998; 26:31.
30. Ryan J. Immersion deaths and swim failure: Implications for resuscitation and prevention. Lancet 1999; 354:613.
31. Ryan T et al. 1999 update: ACC/AHA guidelines for the management of patients with acute myocardial infarction: Executive summary and recommendations. Circulation 1999; 100:1016.
32. Singer A et al. Emergency Medicine PEARLS. Philadelpia: FA Davis, 1996.
33. Soll R. Consensus and controversy over resuscitation of the newborn infant. Lancet 1999; 354:4.
34. Strange G et al. Pediatric Emergency Medicine. New York: McGraw-Hill, 1996.
35. Tapson V. Management of the critically ill patient with pulmonary embolism. J Crit Illness (supplement) July 2000.
36. Tyson J et al. Viability, morbidity and resource use among newborns of 501 to 800 g birth weight: National Institute of Child Health and Human Development Neonatal Research Network. JAMA 1996; 276:1645.
37. Van Hoeyweghen R et al. Quality and efficiency of bystander CPR. Resuscitation 1993; 26:47.
38. Weil M, Tang W. CPR—Resuscitation of the Arrested Heart. Philadelphia: WB Saunders, 1999.

CHAPTER 9

Future and Controversies

Body Temperature and Thermometers

Although temperatures and thermometers are evaluated and scrutinized from time to time, an interesting finding is that the information recorded by Wunderlich in 1871 is quite close to today's data using modern instruments. A recent study by a critic of Wunderlich's work concluded by validating most of what he did. However, the author did find that the mean normal temperature today is 98.2°F (36.8°C) instead of 98.6°F (37°C).

A unique facet of Wunderlich's work was that it was done with a large primitive thermometer held in the axilla, which, with the exception of the neonate, gives the more imprecise of body temperatures today. The reason for Wunderlich's accuracy is the scrupulous manner in which temperatures were recorded (see Chapter 1). An important concept offered by Wunderlich, gradually re-emerging, is that the range of normal temperatures is quite broad and borderline temperatures indicating "fever" are probably irrelevant. Based upon what is known about the hypothalamus and the immune system today, a transient rise in temperature is becoming increasingly apparent as a normal variant.

Which type of thermometer and which location on the body more accurately evaluates a core temperature?

Many variables affect the reliability of an oral reading, and if not done correctly may be in error by several degrees. The rectal thermometer measures the temperature in colic vessels and surrounding tissue distant from the hypothalamus. It thus measures an event that has already taken place, accounting for the erratic reliability of readings. The tympanic thermometer measures temperature from the tympanic cavity adjacent to the hypothalamus, and reflects a core temperature (temperatures measured by an indwelling catheter in the pulmonary artery show a strong correlation with the portable tympanic thermometer). However, the tympanic thermometer is user-dependent. As mentioned in Chapter 2, without a tight seal in the auditory canal the reading may be off by several degrees and the sensor may be partially recording the surface temperature of the auditory canal. A missed fever, particularly in the less than 3-month-of-age group, is critical. If designed differently, the tympanic thermometer could be the instrument of the future. In the meantime, the rectal temperature is probably the more reliable.

Vital Signs and Resuscitation, by Joseph V. Stewart. ©2003 Landes Bioscience.

Four tympanic thermometers are presently available. All but one use "equivalency" settings—that is, presumed oral, rectal and core differences determined by the manufacturer in experimental trials and programmed into the instruments. Unfortunately, no such "equivalency" exists, adding confusion to questionable temperature statistics.

The question may be posed: "Why are we interested in a core temperature at all?" Answer: if the reason is to track a treatment regimen in a critical care unit, an indwelling pulmonary artery or tympanic catheter is appropriate and will give an exact core reading. However, if it is to detect a fever, then we are not interested in a core temperature. The body is merely reacting to the invader and the actual temperature is irrelevant (excluding early pediatrics). A temperature recorded almost anywhere in the body, if performed correctly, will indicate the presence of fever. The key is "performed correctly". An electronic probe placed incorrectly under the tongue or in the rectum while methodically displaying a number may be completely missing a fever. A tympanic probe, placed incorrectly but displaying in two seconds, may miss a high temperature. A glass thermometer, placed correctly in the axilla for the proper amount of time, may correctly identify a fever. The concern, particularly today, is for **accuracy**, not speed.

Heart Rate, Respiration and Blood Pressure

Presently, **oscillometric electronic monitors** that automatically measure blood pressure, heart-rate and the oxygen saturation of blood, are in use in most hospitals. **Portable digital blood pressure monitors** are now available. As with the thermometer, electronic devices are not free from error and the patient may require manual vitals or repositioning, or the electronic device may need recalibration.

Level of Consciousness

At least 40 different "responsiveness" or "coma" scales and scores have materialized over the past 32 years. Bouzarth's watch sheet for brain injury (1968) was revised in 1978 to a cumbersome 100 digital scoring system (see Chapter 1). Over the past few years, attempts have been made to design a simple linear scale. A **Birmingham 9-scale** was actually very simple, but, according to the author, not as sensitive as required. An 8-point linear scoring system (an arousability scale) was introduced in Sweden in 1984 (**Reaction level scale, RLS**), but has not gained wide acceptance (Figs. 9.1, 9.2).

Why does the Glasgow Coma Scale work better than other scales, including a linear system? The key to its effectiveness is simplicity, with the added benefit that each of the components of consciousness is evaluated in three clear subscales. Once learned, it is easy to remember. Once forgotten, it is easy to relearn. Criticisms of the GCS are that,

<u>Levels of Response</u>

9. Alert, rational and fully oriented.
8. Automatism. (Appears fully awake and alert, but gives incorrect information)
7. Drowsy but answers all questions. Mild impairment of orientation.
6. Answers simple questions but confused and irritable, obeys most commands.
5. Answers only "Yes" or "No". Disoriented, restless and confused. Obeys only simplest commands.
4. No obedience to any commands but responds to pain purposefully.
3. No obedience to commands and responds to pain without purpose.
2. Unrousable by any means.
1. Unrousable, no cough reflex and requires artificial respiration.

Fig. 9.1. Levels of Response (Birmingham Accident Hospital).

1. A verbal response cannot be scored in an intubated patient,
2. Eye-opening is difficult to assess in the trauma patient with severe periorbital edema,
3. If drugs are used to intubate the patient, or for other reasons, the scoring system is inaccurate, and
4. The system is imprecise if hypotension is profound.

Answers:

1. Almost without exception, the initial Glasgow Coma Scale evaluation is performed by EMTs in the field before intubation or drugs are given. Some centers use a non-numerical designation of "T" for the verbal score in intubated patients.
2. In our emergency department, we have always been able to evaluate eye-opening, even in the presence of severe facial trauma.
3. When drugs are given (i.e., glucose) or rapid sequence intubation performed (thus involving drugs) a score is assigned before drugs are given.
4. If hypotension and/or hypoxia is profound, this does not negate an initial Glasgow Coma score. When these conditions are corrected, a second score is recorded.

Trauma Scores

Trauma scales appeared shortly after coma scales during the rise of trauma centers in the 1970s. The first of these Injury Severity Scores was the Trauma Index, developed by Kirkpatrick and Youmans in 1971. In 1980, the American Trauma Society under the guidance of H. R. Champion developed the Trauma Score, which included the Glasgow Coma Score, respiratory rate,

1. Alert. No delayed response.
2. Drowsy or confused. Response to light stimulation.
3. Very drowsy or confused. Response to strong stimulation.
4. Unconscious. Localizing but does not ward off pain.
5. Unconscious. Withdrawing movements at pain stimulation.
6. Unconscious. Stereotype flexion movements at pain stimulation.
7. Unconscious. Stereotype extension movements at pain stimulation.
8. Unconscious. No response to pain stimulation.

Fig. 9.2. Reaction Level Scale.

respiratory expansion, systolic blood pressure and capillary refill. Later it was found that capillary refill and respiratory expansion were difficult to assess, particularly at night, and were discarded. This resulted in the **Revised Trauma Score (RTS)** used today in most trauma centers (see Fig. 6.3). In spite of no revisions for two decades, it is still referred to as the "Revised Trauma Score" rather than "Trauma Score", although recently some centers are leaving off the term "revised". Another trauma scoring system, developed in 1982 but not widely utilized, is the **CRAMS Scale** (Circulation, Respiration, Abdomen, Motor, Speech) (Fig. 9.3).

The problem with many scores and scales, such as the Glasgow Outcome Scale—6 months after injury, National Institutes of Health (NIH) Stroke Scale, Cincinnati Prehospital Stroke Scale, Los Angeles Prehospital Stroke Screen (LAPSS), Hunt and Hess Scale (for subarachnoid hemorrhage), Glasgow Meningococcal Septicemia Prognostic Score, Pittsburgh and Ottawa Knee Rules and recently the POGO Score (Percentage Of Glottic Opening for endotracheal intubation) is that unless they are used frequently (major centers) they are not easy to remember, making them impractical for most EMS and emergency departments.

Pediatric Vitals

An ongoing controversial topic is: since the evolutionary reason for fever is to destroy microorganisms, do antipyretics prolong the healing process? Parents and many health professionals when confronted by a child with a fever reach for Tylenol or Advil. Studies often sidestep the issue, and instead discuss whether the response of the infection/fever to antipyretics suggests a life-threatening fever/infection (the results of which during the past ten years have been equivocal and indecisive). By making the child "feel better" with the antipyretic, are we treating the parents (as many suspect we are) and not the child? Conclusive experimental evidence is lacking, but a significant number of studies have suggested that antipyretics do prolong the healing process. (Ref: Bernard, Doran, Graham, Kluger, Mackowiak, Mogabgab, Nielsen,

Smith, Stanley, Whittaker). Studies also support the proposition that antipyretics do not prevent febrile seizures (Ref: Camfield, Nelson, Schnaiderman, Uhari).

Circulation

BP > 100 mm Hg systolic, capillary refill normal 2
BP < 85-100 mm Hg systolic, capillary refill delayed. . . . 1
BP < 85 mm Hg systolic, capillary refill absent 0

Respiration

Normal 2
Abnormal 1
Absent 0

Abdomen

Abdomen, thorax not tender 2
Abdomen or thorax tender 1
Abdomen rigid or flail chest 0

Motor

Normal 2
Responds to pain 1
No response or decerebrate 0

Speech

Normal 2
Confused 1
Unintelligible 0

Score: 0-10
<8: major injury with significant mortality

Fig. 9.3. CRAMS Scale.

Another topic is how fever without a source should be managed in infants. Since *Strep pneumoniae* accounts for 90% of occult bacteremia in the 3-36 month age group today, use of a new pneumococcal vaccine should reduce invasive infections (i.e., sepsis, pneumonia, meningitis) by 90% (Ref: Baraff).

Resuscitation

Recent innovations in resuscitation creating mild controversy are:

1. CPR using chest compressions alone may be as good as CPR with compressions and ventilation (Ref: Berg, Noe, Hallstrom, Van Hoeyweghen),

2. 1-2 minutes of CPR before defibrillation if response times are 4 minutes or longer may improve survival (Ref: Cobb),
3. It may be appropriate, under certain conditions, not to resuscitate (Ref: Krumholz, Goodlin, deVos, AMA Council on Ethical and Judicial Affairs),
4. Withholding resuscitation for newborns with confirmed gestation <23 weeks may be appropriate (Ref: Landwirth, Tyson, Finer).

In 2000, the American Heart Association in collaboration with the International Liaison Committee on Resuscitation (ILCOR) redesigned protocols for Cardiopulmonary Resuscitation (CPR), Advanced Cardiac Life Support (ACLS) and Pediatric Advanced Life Support (PALS). For CPR, the pulse check was discontinued and chest compressions without ventilation were recommended. For ACLS, high-dose epinephrine and bretylium were discontinued. Vasopressin became an acceptable alternative to epinephrine for V-fib and pulseless V-tach. Procainamide and amiodarone (rather than lidocaine) are now first-line agents for V-tach, although amiodarone is expensive and unavailable at many emergency departments. Several algorithms unfortunately became more complex rather than less so and included items that the emergency physician would not know, such as whether a left ventricular ejection fraction was less than or greater than 40% in a patient with atrial fibrillation, flutter or supraventricular tachycardia, and whether the duration of atrial fibrillation or flutter was shorter or longer than 48 hours. Hopefully in the future a more common sense approach will be taken.

Other

One may argue that other parameters such as **abdominal pain** (or the evaluation of pain in general) would be appropriate for consideration as a vital sign. However, as mentioned in the Preface, an abnormality of a vital sign is often life-threatening and must be corrected for survival. Abdominal pain occasionally but not usually falls into this category.

Pulse oximetry has been enthusiastically endorsed by many emergency physicians and pediatricians as a new vital sign. However, because of its solitary frame of reference, it will probably remain as a helpful portion of the "respiration" vital sign.

NOTE: As a nonvital sign note, from time to time a thrust is made to change the term "EKG", the German abbreviation for "elektrokardiogramm", to ECG, the English equivalent. Unfortunately, once enmeshed in the literature, the amending of a medical term or abbreviation is nearly impossible. The same applies to the modern day use of "ED" (emergency department) for the older term "ER" (emergency room), the latter conjuring up for older physicians not the TV series but rather a side room on the 2nd floor of the hospital, where it was located in earlier days.

NOTE: As a last important note, vital signs are rarely static, with the possible exception of temperature. They should be **repeated often** (decreased level of consciousness can sneak up on you). In the very ill, they should be repeated every 15 or 5 minutes.

References

1. Berg R et al. Bystander cardiopulmonary resuscitation: Is ventilation necessary? Circulation 1993; 88:1907.
2. Bernard G et al. The effects of ibuprofen on the physiology and survival of patients with sepsis. N Engl J Med 1997; 336:912.
3. Bouzarth W, Lindermuth J. Head injury watch sheet modified for a digital scale. J Trauma 1978; 18:8.
4. Braakman R et al. Prognosis and prediction of outcome in comatose head injured patients. Acta Neurochir Suppl 1986; 36:112.
5. Camfield R et al. The first febrile seizure—Antipyretic instruction plus either phenobarbital or placebo to prevent recurrence. J Pediatr 1980; 5:719.
6. Champion H et al. A revision of the trauma score. J Trauma 1989; 29:5.
7. Cobb L et al. Influence of cardiopulmonary resuscitation prior to defibrillation in patients with out-of-hospital ventricular fibrillation. JAMA 1999; 281:1182.
8. Communicore Inc. The evolution of clinical thermometry, Newport Beach, 1992, Diatek/Communicore Inc.
9. Council on Ethical and Judicial Affairs, American Medical Association: Guidelines for the appropriate use of do-not- resuscitate orders. JAMA 1991; 265:1868.
10. Finer N et al. Cardiopulmonary resuscitation in the very low birth weight infant: The Vermont Oxford Network experience. Pediatrics 1999; 104:428.
11. deVos R et al. In-hospital cardiopulmonary resuscitation: Prearrest morbidity and outcome. Arch Int Med 1999; 159:845.
12. Doran T et al. Acetaminophen: More harm than good for chickenpox? J Pediatr 1989; 114:1045.
13. Gaddis G and Gaddis M. Non-normality of distribution of glasgow coma scores and Revised trauma scores. Ann Emerg Med 1994; 23:1.
14. Gjerris F. Head injuries in children—Special features. Acta Neurochir Suppl 1986; 36:155.
15. Goodlin S et al. Factors associated with use of cardiopulmonary resuscitation in seriously ill hospitalized adults. JAMA 1999; 282:2333.
16. Graham M et al. Adverse effects of aspirin, acetaminophen, and ibuprofen on immune function, viral shedding, and clinical status in rhinovirus-infected volunteers. J Infect Dis 1990; 162:1277.
17. Hallstrom A et al. Cardiopulmonary resuscitation by chest compression alone or with mouth-to-mouth ventilation. N Engl J Med 2000; 342:1546.
18. Jennett B, Plum F. Persistent vegetative state after brain damage: A syndrome in search of a name. Lancet 1972; 1:734.
19. Jennett B. Clinical assessment of consciousness, Acta Neurochir Suppl 1986; 36:90.
20. Jones I, Champion H. Trauma triage: Vehicle damage as an estimate of injury severity. J Trauma 1989; 29:5.
21. Kluger M et al. The adaptive value of fever. In: Mackowiak P ed. Fever, Basic Mechanisms and Management, 2nd Ed. Philaldelphia: Lippincott/Raven, 1997.
22. Knight R. The Glasgow Coma Scale: Ten years after. Crit Care Nurse 1986; 6:3.
23. Knopp R et al. Mechanism of injury and anatomic injury as criteria for prehospital trauma triage. Ann Emerg Med 1988; 17:9.
24. Krumholz H et al. Resuscitation preferences for heart-failure patients likely to change. Circulation 1998; 98:648.
25. Landwirth J. Ethical issues in pediatric and neonatal resuscitation. Ann Emerg Med 1993; 22:502.
26. Mackowiak P, Plaisance K. Benefits and risks of antipyretic therapy. Ann NY Acad Sci 1998; 856:214.
27. Mackowiak PA, Wasserman SS, Levine MM. A critical appraisal of 98.6°F, the

upper limit of the normal body temperature, and other legacies of Carl Reinhold August Wunderlich. JAMA 1992; 268:1578.

28. Marion D, Carlier P. Problems with initial glasgow coma scale assessment caused by prehospital treatment of patients with head injuries: Results of a national survey. J Trauma 1994; 36:1.

29. Meredith W et al. The conundrum of the Glasgow Coma Scale in intubated patients: A linear regression prediction of the Glasgow verbal score from the Glasgow eye and motor scores. J Trauma 1998; 44:839.

30. Mogabgab W, Pollack R. Increased virus shedding with aspirin treatment of rhinovirus infection. JAMA 1976; 235:801.

31. Nelson K, Ellenberg J. Prognosis in children with febrile seizures. Pediatrics 1978; 61:721.

32. Nielsen V, Webster R. Inhibition of human polymorphonuclear leukocyte functions by ibuprofen. Immunopharmac 1987; 13:61.

33. Nierman D. Core temperature measurement in the intensive care unit. Crit Care Med 1991; 19:818.

34. Noe M et al. Mechanical ventilation may not be essential for initial cardiopulmonary resuscitation. Chest 1995; 108:821.

35. Plaisance K, Mackowiak P. Antipyretic therapy. Physiologic rationale, diagnostic implications, and clinical consequences. Arch Int Med 2000; 160:449.

36. Price D. Factors restricting the use of coma scales. Acta Neurochir Suppl 1986; 36:106.

37. Schnaiderman D et al. Antipyretic effectiveness of acetaminophen in febrile seizures: Ongoing prophylaxis versus sporadic usage. Eur J Pediatr 1993; 152:747.

38. Shinozaki T et al. Infrared tympanic thermometer: Evaluation of a new clinical thermometer. Crit Care Med 1988; 16:148.

39. Smith R. Modulation of phagocytosis by and lysosomal enzyme secretion from guinea-pig neutrophils: Effect of nonsteroid anti-inflammatory agents and prostaglandins. J Pharmacol Exp Ther 1977; 200:647.

40. Sudhammar D, Starmark J. Assessment of responsiveness in head injury patients. Acta Neurochir Suppl 1986; 36:91.

41. Stanley E et al. Increased viral shedding with aspirin treatment of rhinovirus infection. JAMA 1975; 231:1045.

42. Starmark J, Lindgren S. Is it possible to define a general "conscious level"? Acta Neurochir Suppl 1986; 36:103.

43. Sternbach G. The Glasgow Coma Scale. J Emerg Med 2000; 19:67.

44. Stewart JV. Re-evaluation of the tympanic thermometer in the emergency department. Ann Emerg Med 1992; 21:158.

45. Teasdale G, Jennett B. Assessment of coma and impaired consciousness: A practical scale. Lancet 1974; 2:81.

46. Tepas J et al. The pediatric trauma score as a predictor of injury severity in the injured child. J Ped Surg 1987; 22:15.

47. Thomson A et al. Validation of the glasgow meningococcal septicemia prognostic score: A ten-year retrospective survey. Crit Care Med 1991; 19:26.

48. Tyson J et al. Viability, morbidity and resource use among newborns of 501 to 800 g birth weight: National Institute of Child Health and Human Development Neonatal Research Network. JAMA 1996; 276:1645.

49. Uhari M et al. Effect of acetaminophen and of low intermittent doses of diazepam on prevention of recurrences of febrile seizures. J Pediatr 1995; 126:991.

50. Van Hoeyweghen R et al. Quality and efficiency of bystander CPR. Resuscitation 1993; 26:47.

51. Whittaker J et al. The effect of cytotoxic and anti-inflammatory drugs on the phagocytosis of neuatrophil leucocytes. Br J Haematol 1975; 29:273.

SOME CONDITIONS/DRUGS AFFECTING VITAL SIGNS

1. TEMPERATURE:
  Hypothermia:
    Conditions: cold environments, hypoglycemia, adrenal crisis, myxedema, hypopituitarism, sepsis, fluid resuscitation, DKA, extremes of age
    Drugs: alcohol, sedative-hypnotics, drugs of abuse, hypoglycemic agents, opioids
  Hyperthermia:
    Conditions: hot environments, obesity, dehydration, cardiovascular disease, hyperthyroidism, skin diseases, extremes of age
    Drugs: phenothiazines, antihistamines, anticholinergics, diuretics, sympathomimetics (ie cocaine, amphetamines), aspirin

2. PULSE:
  Tachycardia:
    Conditions: emotional stress, pain, exercise, fever, poor cardiopulmonary perfusion (ie MI, CHF, pulmonary embolism, hypovolemia, anemia, sepsis), hyperthyroidism, fast atrial fibrillation, PSVT, cardiac disease (ie ventricular tachycardia), withdrawal, DKA, adrenal crisis, thyroid storm, renal failure
    Drugs: sympathomimetic agents, anticholinergics, alcohol, caffeine, carbon monoxide
  Bradycardia:
    Conditions: physical fitness, cardiac disease (ie MI, AV blocks), increased intracranial pressure, neurogenic shock
    Drugs: opioids, sedative-hypnotics, barbiturates, beta-blockers, calcium-channel blockers, digoxin

3. RESPIRATION:
  Tachypnea:
    Conditions: hypoxia, hypercapnia, acidosis (and combinations, as in asthma, pneumonia, CHF, COPD, pulmonary embolism, pneumothorax), anxiety (ie hyperventilation), obesity, pain, hyperthyroidism, DKA
    Drugs: anticholinergics, sympathomimetics, aspirin overdose, methanol, ethylene glycol poisonings, carbon monoxide
  Bradypnea:
    Conditions: increased intracranial pressure (trauma, hemorrhagic stroke), hypercapnia, many pre-arrest and end-stage conditions
    Drugs: opioids, alcohol, barbiturates, sedative-hypnotics

4. *BLOOD PRESSURE:*
Elevated blood-pressure:
Conditions: behavioral conditions, pain, exercise, poorly controlled hypertension (hypertensive encephalopathy, aortic dissection, cerebrovascular accident), cardiac emergencies (ie MI), secondary hypertension (ie eclampsia, coarctation), alcohol and clonidine withdrawal

Drugs: sympathomimetics (ie cocaine, amphetamines), anticholinergics
Hypotension:
Conditions: shock (ie hypovolemic, cardiogenic, septic, neurogenic, anaphylactic), myxedema, adrenal crisis, DKA, renal failure

Drugs: antihypertensives (ie beta-blockers, calcium-channel blockers), antidepressants, diuretics, opioids, sedative-hypnotics

5. *LEVEL OF CONSCIOUSNESS:*
Agitation:
Conditions: behavioral disorders (psychosis, dementia, delirium), head trauma, meningitis, hyperthermia, thyroid storm, withdrawal, serotonin syndrome, hypoxia, hypoglycemia

Drugs: hallucinogens (ie PCP), sympathomimetics (ie amphetamines, cocaine), alcohol
Lethargy/coma: (TIPS AEIOU)
Conditions: myxedema coma, adrenal crisis, hypoxia, hypercapnia, increased intracranial pressure (ie hemorrhagic CVA, trauma), sepsis, meningitis, shock, DKA, hypoglycemia, uremia

Drugs: alcohol, opioids, sedative-hypnotics, barbiturates, carbon monoxide, aspirin, anticholinergics

A

Abdominal aortic aneurysm 89
Abdominal thrust 134, 135
Abnormal heart sounds 38, 50
Abruptio placentae 92
Acclimatization 23, 29
Acid-base conditions 65
Adrenal insufficiency 110
Advanced Life Support 126, 128, 144, 159
Airway obstruction 67, 134, 147
Alcohol 1, 4, 28, 66, 81, 84, 86, 96, 101, 105, 109, 123-125, 152
ALTE 81, 82, 85, 89, 97, 100, 107, 110, 111, 116, 120, 123, 144, 151, 159
Anaphylaxis 67, 91
Aneroid manometer 77
Angioedema 67
Antipyretics 26, 27, 157, 158
Aortic aneurysm 50, 51, 89
Aortic dissection 81, 82
APGAR score 113
Apparent life-threatening event 116
Arrhythmias 30, 39, 41, 43, 89, 116, 151, 152
Asthma 58, 59, 64, 65, 68, 70, 80, 111, 119, 120, 132, 136
Asystole 138, 141, 143, 149, 151
Atrial fibrillation 39, 41, 43, 45, 47, 51, 159
Atrial flutter 41, 43
Auscultation of the heart 89
Axillary temperture 5, 23, 24, 32, 136, 146

B

Bacteremia 90, 107, 158
Bacterial meningitis 106
Baroreceptors 40, 62, 52, 81, 84, 86, 88, 101
Basic Life Support 128, 144
Beta-blocking agents 38
Blood gas analyzers 61
Blood gases 68, 69, 119
Blood-pressure 52, 77, 78, 80, 86, 87, 120

Boerhaave, Herman 4, 14
Boyle, Robert 2
Brachial artery 53, 74, 77, 78
Brachial pulse 56, 78
Bradycardia 34, 39, 45, 46, 49, 85, 86, 89, 91, 101, 110, 119, 128, 138
Bradypnea 65
Brainstem ischemic stroke 107
Bronchiolitis 120
Brudzinski sign 107

C

Calcium-channel blocking agents 35
Capillary refill 82, 84, 88, 121, 122, 126, 157
Carbon monoxide 110, 151
Cardiac arrest 30, 85, 114, 151
Cardiac arrhythmia 30
Cardiac tamponade 52, 81, 84, 89, 90, 105, 137
Cardiogenic shock 89, 90
Carotid artery 12, 40, 74
Carotid massage 40
Causes of coma 105, 107
Celsius 4
Chemoreceptor 61, 62, 84
Chest compressions 128, 131, 144, 148, 150, 159
Chest tube 58, 131, 136, 137, 146
Cheyne-Stokes 66
Chilblain 30
Choking 116, 118, 134
Click 50
Coarctation of the aorta 83
Coma 10, 17, 25, 56, 65, 96, 99, 101, 102, 105, 107, 109, 110, 112, 123, 125, 128, 132, 136, 145, 149, 155, 156
Coma scales 156
Common carotid artery 52
Common cause of a seizure 108
Concussion 106
Conducting system of the heart 36
Contusion 106, 136
COPD 52, 58, 64, 65, 67, 68, 80
Core temperature 23, 26, 29, 30, 115, 154, 155

Cortex 21, 61, 83, 96
Cortical ischemic stroke 107
Cricothyrotomy 58, 67, 133, 134, 135, 145, 146
Croup 119
Cushing reflex 101

D

Decerebrate posturing 103
Decorticate posturing 103, 105
Delirium 96
Dementia 96
Dependency 109
Diabetic ketoacidosis 62, 65, 66, 105, 111, 126
Dialysis 83, 93, 112, 151
Diffuse axonal injury 106
Disability 58, 131, 140, 149
DKA 111
Doppler 49, 50, 56, 70, 78, 120
Dorsal pedis artery 55
Dyspnea 42, 67, 69, 70, 72

E

Ectopic pregnancy 85, 92
Electrical injuries 151
Electrolyte 109, 111, 125
Emphysema 59, 68
Epidural hematoma 106
Epiglottitis 119

F

Fahrenheit, G.D. 4, 5, 14
Febrile seizures 115, 158
Femoral 54-56, 78-80
Fetal heart tones (FHTs) 50
Fever 1, 4-6, 15, 26, 27, 32, 39, 69, 80, 107, 114-116, 119, 120, 122-124, 126, 154, 155, 157, 158
Friction rub 50
Frostbite 30

G

GABA receptors 110
Galileo 1, 2, 8
Glasgow Coma Scale 17

H

Hales, Stephen 12
Heart murmur 50
Heart-rate 34-36, 49, 85-88, 97, 113, 150, 155
Heat cramps 29
Heat exhaustion 29
Heat stroke 26, 27, 29, 30
Heimlich maneuver 134, 135, 145
Hemorrhagic nontachycardia 85
Hemorrhagic stroke 101, 107, 123
Hepatic encephalopathy 109
Hooke, Robert 2, 11
Hypercapnia 69, 111, 137
Hyperkalemia 110, 111, 151
Hyperosmolar hyperglycemic nonketotic coma 111
Hypertensive 66, 81-84, 104, 137
Hypertensive encephalopathy 66, 81
Hyperthermia 26, 28, 29, 32, 103, 106, 151
Hyperventilation 32, 65, 91, 101, 104
Hypoglycemia 30, 96, 101, 103, 111, 123, 126
Hypotension 32, 40, 45, 70, 75, 79, 84, 85, 87, 89, 91-93, 104, 107, 110, 122, 156
Hypothalamus 20, 21, 23, 26, 97, 154
Hypothermia 30, 31, 32, 58
Hypothyroidism 30, 103, 110
Hypovolemic shock from blood loss 121

I

In-line immobilization 133
Indirect monitoring 75
Intracranial pressure 66, 80, 82, 103, 106, 108, 123, 133, 140, 149
Intraosseous access 148

J

Jennett, Bryan 17

K

Kernig sign 107
Korotkoff 12
Kussmaul 9

L

Laennec, Rene 8
Lightning injuries 151
Lung volumes 11, 59

M

Metabolic acidosis 63, 65, 66, 91, 114
Metabolic alkalosis 63, 65
Monoamine oxidase inhibitor
    antidepressants 83
Multifocal atrial tachycardia 48

N

Nasotracheal intubation 132, 145
Near-drowning 152
Needle cricothyrotomy 133, 145, 146
Needle thoracentesis 131, 136
Neonatal resuscitation 150, 151
Neonate 114, 115, 123, 154
Neurogenic shock 91
Neuroleptic malignant syndrome 29,
    32

O

Oculovestibular reflex 103, 105
Orotracheal intubation 132
Orthostatic vital signs 49, 57, 85, 86,
    121
Oscillometric electronic monitors 155
Oxygen saturation 61, 70, 75, 111,
    120, 155
Oxyhemoglobin dissociation curve 61

P

Paradoxical bradycardia 49, 86
Paradoxical pulse 51, 80
Paroxysmal supraventricular tachycar-
    dia (PSVTs) 39, 41
Partial obstruction 118, 119, 134, 135

Peak flow 119, 120
Pediatric advanced life support 126,
    144, 159
Pediatric Basic Life Support 144
Pediatric fever 114
Pediatric Glasgow Coma Scale 123,
    125, 149
Pediatric trauma score 148, 149
Percutaneous transluminal coronary
    angioplasty 90
Pericardial tamponade 105
Peripheral pulses 52
Pheochromocytoma 83
Placenta previa 92
Pneumonia 64, 65, 67, 69
Pneumothorax 58, 65-67, 70, 90,
    105, 136, 145, 146
Portable digital blood pressure
    monitors 155
Posterior tibial artery 54, 56, 78
Postpartum hemorrhage 92, 93
Postural orthostatic tachycardia
    syndrome 87
Pre-eclampsia 83
Pregnancy (cardiac arrest) 151
Premature ventricular contractions 39,
    47, 51
Pulmonary edema 58, 67, 69, 70
Pulmonary embolism 65, 67, 70
Pulmonary Function Testing 59
Pulse 5, 7-9, 12, 13, 15, 30, 34, 38,
    39, 40, 42, 49, 51-54, 56, 58, 60,
    61, 68, 69, 78, 79, 81, 82, 84, 86,
    87, 89, 90-93, 111, 113, 116, 119,
    120, 122, 128, 131, 133, 137-141,
    144, 149, 159
Pulse deficit 51, 56
Pulse oximeter 61
Pulse oximetry 58, 68, 89, 92, 111,
    119, 120, 159
Pulse pressure 75, 81, 84, 86, 91, 122
Pulseless electrical activity (PEAs) 77,
    78, 80, 81, 87, 93, 113, 116, 119,
    120, 134, 138, 140, 141, 149, 157
Pulseless ventricular tachycardia 137,
    138, 139, 149
Pulsus alternans 51
Pulsus paradoxus 51, 52, 80, 90

R

Radial 51-53, 56, 75, 77-80
Rales 66, 69, 71, 89
Rapid sequence intubation 125, 133, 145, 156
Reaction level scale 155, 156
Rectal 5, 6, 23, 24, 30, 32, 114, 115, 137, 154, 155
Rectal temperature 23, 24, 30, 109, 114, 115, 154
Relative bradycardia 85
Renovascular hypertension 83
Respiratory acidosis 64, 65, 111
Respiratory alkalosis 65, 68
Respiratory center 61, 62, 65
Respiratory failure 32, 68, 120, 133, 136, 147
Reticular activating system 97
Revised Trauma Score 98, 123, 124, 149, 157
Riva-Rocci 125

S

Secondary hypertension 83
Seizure 83, 108, 115, 116, 152
Sepsis 30, 32, 65, 90, 103, 107, 115, 116, 122, 158
Septic shock 90, 107, 122, 124
Severe headache 108
Shock 79, 81, 84, 85, 89-92, 104, 105, 107, 108, 114, 120-124, 126, 137, 149, 151
Sinus arrhythmia 45, 47
Sinus tachycardia 39, 41, 116
Sleep apnea 66
Sounds of Korotkoff 77
Splitting of the 1st sound 50
Splitting of the 2nd sound 50
Stridor 119, 128, 134, 147
Stroke 8, 26, 27, 29, 30, 40, 51, 66, 81, 96, 97, 101, 108, 123, 157

Subarachnoid hemorrhage 82, 106, 108, 157
Subdural hematoma 106, 109
Sudden infant death syndrome (SIDS) 116
Supraventricular tachycardia 39, 41, 43, 116, 159
Systolic pressure 12, 49, 69, 76, 77-80, 82, 84-86, 89, 91, 101, 120

T

Tachycardia 28, 29, 34, 39, 40, 41, 43, 44, 48, 49, 69, 70, 115, 116, 120, 122, 126
Tachypnea 65-67, 69, 70
Teasdale, Graham 17, 98
Tension pneumothorax 58, 70, 90, 105, 136, 145
Tilt test 87
Tolerance 109
Trauma scores 156
Tympanic temperature 30, 32

U

Uremia 66, 104, 111

V

Vaginal temperature 24
Ventricular fibrillation 137, 139, 140, 149
Ventricular tachycardia 39, 41, 43, 44, 116, 137-139, 149, 159
von Basch 12

W

Wheezes 66, 68, 69, 72
Wunderlich 5-7, 15